THE PLANTATION SOUTH

TOURING NORTH AMERICA

SERIES EDITOR
Anthony R. de Souza, *National Geographic Society*

MANAGING EDITOR
Winfield Swanson, *National Geographic Society*

THE PLANTATION SOUTH

*Atlanta to Savannah
and Charleston*

BY
LOUIS De VORSEY, Jr.
AND
MARION J. RICE

RUTGERS UNIVERSITY PRESS • NEW BRUNSWICK, NEW JERSEY

This book is published in cooperation with the 27th International Geographical Congress, which is the sole sponsor of *Touring North America*. The book has been brought to publication with the generous assistance of a grant from the National Science Foundation/Education and Human Resources, Washington, D.C.

Copyright © 1992 by the 27th International Geographical Congress
All rights reserved
Manufactured in the United States of America

Rutgers University Press
109 Church Street
New Brunswick, New Jersey

The paper used in this book meets the minimum requirements of American National Standard for Information Sciences—Permanence of Paper for Printed Library Materials, ANSI Z39.48-1984.

Library of Congress Cataloging-in-Publocation Data
De Vorsey, Louis.
  The plantation South: Atlanta to Savannah and Charleston / by Louis De Vorsey, Jr. and Marion J. Rice.
    p.   cm. —(Touring North America)
  Includes bibliographical references and index.
  ISBN 0-8135-1872-5—ISBN 0-8135-1873-3 (pbk.)
  1. Georgia—Tours. 2. South Carolina—Tours. I. Rice, Marion J. II. Title.
  III. Series.
  F284.3.D4   1992
  917.58—dc20   92-10413
                                                                                CIP
First Edition

Frontispiece: Cotton in bloom, south Georgia. Photograph by Lynda Sterling.

The cover photograph is of Casulon Plantation, 1824, in the heart of the Old Cotton Belt. The former home of Georgia governor James Boynton, it is listed on the National Register of Historic Places. The house is noted for its 12-star door frame corner blocks, Adam mantels, and faux wainscoting, unusual in a country home of the period; the grounds contain a boxwood parterre garden, stone wall–enclosed kitchen garden, and original outbuildings. Now a private residence, there is a good view of the house and grounds from the Jones Woods Road near the intersection with Ga. 186, southeast of Monroe and southwest of Athens; see Bishop, p. 48 text. Group tours, weddings, and receptions may be arranged through LIFE (Local Individuals for Environment) Group, Inc., 5784 Lake Forrest Drive, Suite 204, Atlanta, Ga. 30328, Janice Sommer, (404) 252-1000; proceeds to protect an environmentally threatened habitat in Jones Woods area.

Series design by John Romer

Typeset by Peter Strupp/Princeton Editorial Associates

Contents

FOREWORD ix
ACKNOWLEDGMENTS xi

PART ONE

INTRODUCTION TO THE REGION

Genesis and Evolution of the Plantation in
   Georgia and South Carolina 3
The Transition to Contemporary Agriculture 9
Circle and Square: Post-Revolutionary Order
   Inscribed on the Landscape 15

PART TWO

THE ITINERARY

Highlights of the Region 25
DAY ONE: Atlanta, Georgia 28
DAY TWO: Atlanta to Athens, Georgia 38
DAY THREE: Athens to Macon, Georgia 42
DAY FOUR: Macon and Vicinity 71
DAY FIVE: Macon to Tifton, Georgia 83
DAY SIX: Tifton to Jekyll Island, Georgia 95
DAY SEVEN: Jekyll Island and Vicinity 107
DAY EIGHT: St. Simons Island to Savannah, Georgia 127
DAY NINE: Savannah and Vicinity 133
DAY TEN: Savannah, Georgia, to
   Charleston, South Carolina 149
DAY ELEVEN: Charleston and Vicinity 163

PART THREE
RESOURCES

Hints to the Traveler	*181*
Suggested Readings	*186*
INDEX	*189*

◬ Foreword

Touring North America is a series of field guides by leading professional authorities under the auspices of the 1992 International Geographical Congress. These meetings of the International Geographical Union (IGU) have convened every four years for over a century. Field guides of the IGU have become established as significant scholarly contributions to the literature of field analysis. Their significance is that they relate field facts to conceptual frameworks.

Unlike the last Congress in the United States in 1952, which had only four field seminars, the 1992 IGC entails 13 field guides ranging from the low latitudes of the Caribbean to the polar regions of Canada, and from the prehistoric relics of pre-Columbian Mexico to the contemporary megalopolitan eastern United States. This series also continues the tradition of a transcontinental traverse from the nation's capital to the California coast.

The Plantation South reveals the rich tapestry from prehistoric times through colonial settlement, the Confederacy, to the current period. Examined are archaeological sites, the rice coast cities of Georgia and South Carolina, the historic naval stores industry, the Southern agricultural enterprise from the late eighteenth century to the recent past, the heritage of the antebellum cotton belt to the social turmoil under the leadership of Martin Luther King, Jr. The threads of this fabric are interwoven by Professors Louis De Vorsey and Marion J. Rice of the University of Georgia, longtime scholars of the historical geography of the American South.

Anthony R. de Souza
BETHESDA, MARYLAND

△ Acknowledgments

We acknowledge the dedicated work of the following cartographic interns at the National Geographic Society, who were responsible for producing the maps that appear in this book: Nikolas H. Huffman, cartographic designer for the 27th IGC; Patrick Gaul, GIS specialist at COMSIS in Sacramento, California: Scott Oglesby, who was reponsible for the relief artwork; Lynda Barker; Michael B. Shirreffs; and Alisa Pengue Solomon. Assistance was provided by the staff at the National Geographic Society, especially the Map Library and Book Collection, the Cartographic Division, Computer Applications, Typographic Services, and Illustrations Library. Special thanks go to Susie Friedman of Computer Applications for procuring the hardware needed to complete this project on schedule.

We thank Lynda Sterling, publicity manager and executive assistant to Anthony R. de Souza, the series editor; Richard Walker, editorial assistant at the 27th International Geographical Congress; Elizabeth W. Fisher, copyeditor; Natalie Jacobus, proofreader; and Tod Sukontarak, indexer and photo researcher. They were major players behind the scenes. Many thanks, also, to all those at Rutgers University Press who had a hand in the making of this book, especially Kenneth Arnold, Karen Reeds, Marilyn Campbell, and Barbara Kopel.

Errors of fact, omission, or interpretation are entirely our responsibility; the opinions and interpretations are not necessarily those of the 27th International Geographical Congress, which is the sponsor of this field guide and the *Touring North America* series.

PART ONE

Introduction to the Region

GENESIS AND EVOLUTION OF ⚠ THE PLANTATION IN ⚠ GEORGIA AND ⚠ SOUTH CAROLINA

South Carolina (1670) and Georgia (1733) both began as British settlements when mercantilist thought dominated the economics and politics of settlement. Colonies were designed to serve the interests of the home country, providing needed raw materials and serving as outlets for an excess or undesirable population. The processing and manufacturing of commodities, as well as the carrying trade, were reserved for residents of the mother country.

Mercantilist thought envisioned a flow of profit from colonial peripheries to the core; but it was sufficiently flexible to permit the subsidy of colonial produce by high import duties on the goods of other nations. In both Georgia and South Carolina, where the British subsidy was crucial to the prosperity of indigo growing, the growth of this plantation staple collapsed with the Revolution.

Under British preferential trade, both South Carolina and Georgia found in the established Caribbean Islands an outlet for some of their earliest produce—naval stores and timber, salted beef and pork, and oranges. Another South Carolina commodity, touched upon ever so gently in historical accounts, was the active pursuit and export of Indian slaves to Caribbean purchasers.

The idea of plantation slavery for profit, important in the development of South Carolina and Georgia, was well established. The practice of slavery was alien neither to the lords proprietor nor to the Barbadian colonists who made up a large part of the first South Carolina settlement; but little attention has been paid to the proprietors' economic interests or the Barbadian connection. Of the eight lords proprietor, seven were either planters experienced with the use of African slaves or connected in some way with the African slave trade. The exception was the Duke of Albemarle, and even he had correspondents in Barbados.

It is thus not surprising that slavery was deliberately introduced into South Carolina. By the time the colony was established, English planters in Barbados and Jamaica had already become rich from sugar cane grown with the sweat of black slaves. Seasoned slaves could be brought in with their owners under favorable headright land allotments, and were also available for import. The problem was to find suitable staples for intensive cultivation. Until indigo and rice emerged, black labor was used to exploit resources at hand for the taking—the rich forests and extensive grazing lands suitable for cattle and swine.

The design of the Georgia colony was vastly different. Carved from the original Carolina grant, by then a royal colony, Georgia stemmed from the charitable impulses of the Reverend Dr. Thomas Bray and his Anglican associates and from the need for a frontier buffer against the Spanish to the south. When Georgia came into existence over fifty years after the first English settlement in South Carolina, the Georgia trustees prohibited slavery and limited the size of land grants. This seemed to contradict the economic foundations of its older and more established neighbor. However, James Oglethorpe was far more concerned with the Spanish military question than in guiding Georgia into a Utopia based on principles of independent yeomanry. Slavery had already been illegally introduced into Georgia when the trustees surrendered their twenty-one-year charter; and it expanded rapidly after Georgia became a royal colony. Slavery seemed to provide ambitious men of a preindustrial economy with a ready source of wealth. It became entrenched with the expansion of cotton production in the early nineteenth century at a time when the rest of the United States and Europe were repudiating slavery on moral grounds and experiencing an industrial revolution.

The rush to profit from slavery had depressing economic, political, and moral consequences long after the Civil War had technically ended the slave-plantation system. Economically, a relatively inefficient system of peasant production replaced the centralized plantation from 1865 to 1950; politically, the South continued in regional isolation; and morally, a segregated society replaced slavery.

Changes in textile fashion and technology were associated with the emergence of cotton as the quintessential crop of the Plantation South. The usual technical explanation is that the invention of the gin by Eli Whitney in 1793 (with saw improvements by Hogden Holmes in 1796) made possible an upland-cotton economy and created a new demand for slaves. The story is more complicated, however, than this simple linear explanation of Southern planter response to Yankee innovation. Had there been no demand for cotton, the invention of the saw gin would have been a useful but hardly revolutionary device.

However, all the elements were in place for a factory-based cotton industry, including a strong demand for a great variety of cotton fabrics, from coarse osnaburgs to fine muslins. In the United States the missing supply-side aspect came with the invention of a gin that could remove the seeds from upland, short-staple cotton. This plant could be grown under a greater variety of soil and climatic conditions than the Sea Island variety of cotton. In 1790, before the invention of the saw gin, the world trade of cotton was estimated at only fifty thousand bales, with another million bales used locally. South Carolina and Georgia cotton production in 1790, mainly Sea Island, was 3,000 bales; twenty years later, it was 170,000 bales, mainly upland. Each succeeding decade saw more land in cotton and more slaves as the cotton kingdom reached westward to embrace Texas and Arkansas.

The soils and climate of the South were suited to a plantation economy. The area had long, hot summers and could produce a variety of crops not suitable for cultivation in the higher latitudes with cooler temperatures and shorter growing seasons. The Georgia-Carolina growing season—270 days frost-free along the coast to 220 days in the upper Piedmont—allowed more than ample time to plant, grow, and harvest rice, cotton, and indigo. The growing season was too short, however, for the development of a sugar cane industry.

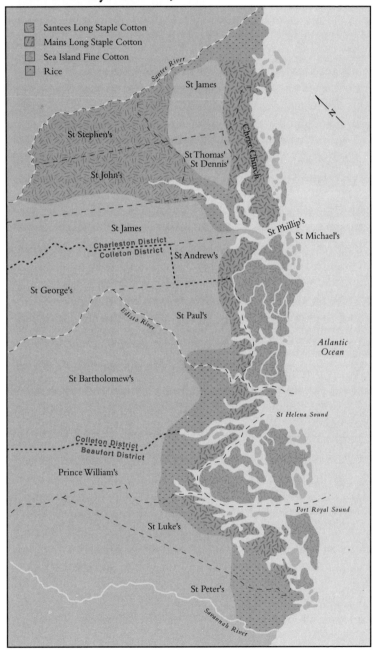

The prevailing technology largely restricted rice growing to the tidal estuaries. The Sea Island variety of long-staple cotton was originally thought to be restricted, by moisture content of the air, to coastal lands. Its diffusion into a lower Coastal Plain west of the Okeefenokee in Georgia suggests that growth was less limited by climate than by tradition and the availability of a labor force accustomed to the demands of this type of cultivation. Upland cotton, a hardier and less sensitive plant, could be grown on the heavy clay soils of the Piedmont as well as the sandier soils of the Coastal Plain which are poorer in nutrients.

Despite the lushness of low-country vegetation, settlers in the Coastal Plain found that without the application of fertilizer, soil exhaustion was common after the second crop. Therefore, eighteenth- and early nineteenth-century settlers preferred the heavier Piedmont soils. Before the topsoil was eroded to reveal the red soils underneath, they were reported as rich, brown soils. The climax vegetation of the Piedmont was deciduous hardwood, a forest type conventionally associated with better soils, whereas the sandy, lower Coastal Plain was often contemptuously called the "pine barrens."

Rainfall in Georgia and South Carolina varies from forty to fifty inches a year over most of the area, with ten-inch higher rainfall in the mountains. Rainfall is distributed throughout the year; but summer temperatures fall in the 90 degrees Fahrenheit range. By the end of the summer, most of the area shows a water deficit, and it takes tropical hurricanes in the fall to bring in sufficient rain to restore a good water supply. When these wind-brought rains fail, there can be a sharp drop in the lake levels and a failure to recharge the aquifiers on which most Georgia and South Carolina farmers rely for pumped irrigation. Microclimatic variations occur, with sometimes devastating effects on farmers in small areas that get too much or too little rain. Generally August is a month of lesser rainfall, especially favorable to harvesting cotton in the upper Coastal Plain. Piedmont harvest generally came a month later, in September or October, since cotton planting was delayed until late April after the danger of late spring frosts was past.

8 THE PLANTATION SOUTH

Far more important than climate or soils to the production of cotton, which moved westward with the slave-holding population, was the availability of cheap land after the United States extinguished Indian claims to territory in the South. The South Carolinians had effectively extinguished an Indian presence in the state by the time of the Revolution; Georgia was in constant conflict with the national government over Indian lands until the final expulsion of the Indians in the 1830s. (A mountain topography made most of the Cherokee country more suitable for small subsistence farms than for plantations, so slavery never got a strong foothold here. Even today some Georgia mountain counties have no black population.)

On the whole, world economic, industrial, and social conditions had far more effect on the emergence of a Plantation South than did the specifics of any particular environment.

THE TRANSITION TO  CONTEMPORARY AGRICULTURE

The stereotypic image of the Plantation South is epitomized by the novel *Gone With the Wind*. It begins with a scene on the veranda at Tara, the red fields of the Piedmont, fine cotton land, making a rich contrast with varied greens of lawn and forest. In the distance, crowning a hilltop, stands the many-columned mansion Twelve Oaks, seat of the Wilkeses, a family distinguished from its cruder north Georgia neighbors by wealth, education, and culture.

Margaret Mitchell, in her delineation of the prevailing social structure, makes it clear that small and large planters existed in a sea of blacks and also in a larger white society that owned no slaves—yeoman farmers, laborers, hunters, artisans, and hangers-on. This latter group has often been overlooked when describing the complex society of the Old South. Too often historians and geographers, following the lead of historian Ulrich B. Phillips of Georgia, focused on life on large plantations and tended to neglect the plain people who were the majority of the whites.

The Plantation South did not die at the end of the Civil War, but it changed in a number of ways. The most dramatic change was that capitalized labor, in the form of slavery, was replaced with wage labor and sharecropping. The abolition of slavery made African Americans free in name; but socially and politically the results were less positive. When the Republican Party abandoned the freedman with the Hayes–Tilden compromise of 1876, it also abandoned enforcement of the Civil Rights Act of 1866 (a situation that lasted for almost a hundred years—until the Civil Rights Act of 1964). The result was a segregated, Jim Crow society in which blacks continued to be the proverbial hewers of wood and drawers of water. Gradually disenfranchised and deprived of civil rights, blacks saw their economic status likewise decline. Under the post–Civil War plantation system, they worked for little more than room and board—a remuneration, whether paid as cash wage or as a share of crops, akin to that under slavery. The effects were disastrous for whites also: More and more, formerly independent farm owners slipped into tenancy and sharecropper status, with income, living standards, and education scarcely above that of blacks.

One of the earliest accounts of the economic and social devastation that accompanied the collapse of the cotton economy is described in *The Souls of Black Folk* (1903) by sociologist-historian W.E.B. DuBois. He collected data in Dougherty County, Georgia, which this traverse goes through. In 1979 79 percent of the county's black population fell below the federal threshold poverty level (then $7,412 for a family of four).

The depressed state of the Plantation South reflected the devastating effect of the Civil War on the capital infrastructure. The area had always been a debtor region: to England in the colonial period and then to the Northeast before and after the Civil War. Its major asset was capitalized labor, in the form of slave ownership. Freeing slaves without compensation reduced personal estate assessments in 1870 in Georgia and South Carolina to 18.5 percent of the 1860 value.

In 1871, Georgia banks had only a little over $2 million in deposits compared to $13.5 million in 1860. The industrialization of the New South did not offset population growth. In the sixty-year period 1849–1909, Georgia manufacturing increased in value from $7 million to $203 million; but, as a proportion of U.S. manufactures, the gain was only three-tenths of one percent. With cotton selling at six cents a pound in the 1890s, less than the cost of production, there was both underemployment and unemployment on the farm as well as in the cities. In the 1930s, in the midst of a worldwide depression, President Franklin D. Roosevelt declared the South to be the nation's number-one economic problem. Per capita farm income there, for both blacks and whites, was less than $200 a year. The South was a third-world region of an industrialized country, long before the concept of a "third world" was formalized.

At that time, the South was still a region of farms and farm people, and quiet country towns of Georgia and South Carolina became rural metropolises on Saturdays as farmers flocked to towns in buggies, wagons, and jalopies to talk, trade, gawk, and buy. There was a nascent vitality in the farmscapes of rustling corn and blossoming cotton, despite the inevitable vicissitudes of too little or too much rain and low farm prices only partially offset by a complex price support system. With a "live at home" family farm philosophy sponsored by farm organizations and the U.S. Department of Agriculture, many farmers prided themselves on their self-sufficiency. With good crop rotation and favorable seasons, some farmers managed to make a bale of cotton to the acre.

The farm people were most important in giving character to the land. In a one-mule plow culture, a veritable army of hands was needed to cultivate row crops, to thin and weed cotton with a hoe—the work of women in the 1930s as it had been a hundred years before—and to provide the nimble fingers to pull the cotton wool from the clutching boll. As late as 1940, 45 percent of the total population of Georgia and South Carolina was classified as "farm"; in 1980, the latest year for which data are available, the figures were 4 and 3 percent, respectively. Farm income now plays an insignificant role in personal income. In Georgia, in 1988, it was under 2 percent; South Carolina, in 1989, less than 1 percent.

The speeding traveler who leaves the interstate highways in South Carolina or Georgia will be hard pressed to find evidence of a Plantation South. Thousands of acres of former cropland lie abandoned to weeds; and in the former cotton-producing Piedmont, loblolly pine is the major crop—planted like cabbages in rows to be harvested with machines. A drive through the country reveals the desolation of abandoned farm and tenant buildings, crumbling in decay or covered in summer with the embrace of kudzu; of empty farm hamlets that once functioned as local service centers with bank, stores, warehouse, and depot; of towns dead on

Saturday with a fading movie-theater sign speaking of happier days; of rural populations composed of the young and the old, but with few workers in their productive years.

In South Carolina and Georgia, the areas with most agricultural productivity follow the upper Coastal Plain, with an area projecting into the Dougherty Plain and Tifton Upland soil provinces of southwest Georgia. On this traverse, the most visible evidence of contemporary, mechanized factories in the field are found south of Macon and on to Albany and Tifton. This area in southwest Georgia accounts for most of the state's crop production, including cotton.

The year 1988 was the biggest cotton year ever in the United States, with the production of over 15 million bales. But of the former "million bale" cotton states east of the Mississippi, only the rich cotton lands of the Mississippi delta held up in production. Georgia and South Carolina, which produced over 10 percent of world supply in 1920, only produced 510,000 bales in 1988, less than one percent of the world supply. Texas, California, and other southwestern states are now the big cotton producers; but the U.S. proportion of world production has declined, accounting for only 18 percent in 1988 compared to 80 percent in 1860.

Rice is another staple that has shifted from Georgia and South Carolina to the west. Once the "Rice Coast," from Cape Fear, North Carolina, to St. Mary's, Georgia, produced almost all the rice grown in the United States—using a technique adapted from West Africa. A series of killer hurricanes at the end of the nineteenth century and the competition of upland rice grown in Arkansas, Louisiana, Texas, and California led to the cessation of cultivation.

Cotton and rice have been replaced in South Carolina and Georgia by two crops: winter wheat and soybeans. These are planted in succession, where a long growing season permits the harvesting of two sequential crops. In 1989 it was expected that this combination would account for nearly 15 percent of farm receipts in South Carolina, closely followed by tobacco with over 14 percent. Soybean yield in both states is hampered by a soil that is moisture deficient in the bean-forming stage in late summer after months of high evapotranspiration—a climate regime more suitable to cotton. Especially important to the farm economy of Georgia is peanuts, brought to national attention by former president Jimmy Carter, who came from a peanut-farming tradition. The big gainers in farm receipts in both South Carolina and Georgia have been livestock and poultry products, which in 1989 accounted for nearly 45 percent of South Carolina farm receipts. The broiler industry, with long, mechanized poultry houses, is concentrated in the northern parts of both states for climatological reasons. During a heat wave, three or four degrees may mean the difference between safety and severe losses from overheating. Sheep, important in the livestock economy of Georgia and South Carolina before the Civil War for both meat and wool, have disappeared.

It is doubtful if the mechanized farm economy of the 1990s is as healthy overall as the farm economy of the 1860s, with its mix of large and small farms, yeoman farmers and large planters, and large inventory of working and food animals. Many problems of the farm economy in the South today are related to the higher ratio of production expenses—machinery, petroleum products, commercial fertilizer, and agricultural chemicals—all requiring a high cash flow to the nonfarm economy.

Ever since the replacement of a multiplicity of craft suppliers of agricultural needs with a few industrial suppliers, agriculture has been in difficulty. This phenomenon is worldwide. In Europe especially, the problem of farm maintenance has been met by agricultural subsidies. In the United States, there are also agricultural subsidies tied to the concept of parity, an ideal price that should maintain farm income at a pre–World War I level. However, the idea of subsidy has been constantly under attack, especially since its most obvious benefits have gone to large commercial farmers rather than those on smaller, family farms.

With the ability of third world countries to grow crops that have long been a mainstay of Southern production—corn, cotton, peanuts, rice, and soybeans—the future of agriculture in Georgia and South Carolina is uncertain. Since few young people are entering agriculture, it is likely that agriculture will become even more concentrated, with larger and fewer units of production—factories in the field with no cultural or moral roots to the land where they operate. Agriculture, whether viewed from the perspective of a plantation or yeoman antebellum South, will then have finally "gone with the wind."

CIRCLE AND SQUARE:
△ POST-REVOLUTIONARY
△ ORDER INSCRIBED ON
△ THE LANDSCAPE

Seen from the air, the landscape of the Plantation South resembles a palimpsest inscribed with the line and pattern signatures of countless generations of red, white, and black inhabitants. Were the traveler flying over this transect, rather than earthbound in an automobile, a rich historical tapestry would unfold below. Without access to high overlooks, aerial photographs, or detailed maps, a traveler may be unaware of the nature and richness of the landscape's designs and patterns.

Over the bulk of the Plantation South, as with the United States as a whole, the primary patterns comprising the landscape are almost invariably those connected with human occupation. Geographer Norman J. G. Thrower addressed this in the introduction to his 1966 monograph *Original Survey and Land Subdivision*.

> Inscribed upon that grand Design, the surface of the earth, are the marks of human occupance. Patterns resulting from man's activities, although individually not of the great scale of some natural features, in aggregate give to certain areas their most distinctive character. Of all the works of man, one of the most widespread, if not the most important, is the subdivision of land. Once laid down, the boundaries of land divisions become part of man's inheritance to be accepted or modified by later generations.

The inherited boundaries of land divisions form an indelible and enduring landscape geometry. It is a geometry influencing individual and societal land-use decision making at every turn. Wolfgang Langewiesche, a German aviator and frequent contributor to *Harper's* magazine, drew attention to the familiar checkerboard pattern of rectangular township and range lines and farm fields that dominate most of the United States from the Appalachians to the Rockies in his 1940 essay "The U.S.A. from the Air." Describing a flight across the United States, Langewiesche wrote that the Middle-Western section lines were "one of the odd sights of the world, and it is strictly an air-sight: a whole country laid out in a mathematical gridwork in sections one mile square each; exact, straight-sided, lined up in endless lanes that run precisely—and I mean precisely—north-south and east-west. It makes the country look like a giant real-estate development: which it is."

Had the German aviator set his flight path across the Plantation South he would have seen an even more remarkable "air-sight" unfolding on the ground below. Over most of South Carolina and the eastern third of Georgia, areas settled when colonial authority held sway, the pilot would behold a ragged, patchwork quilt created by what has been termed "indiscriminate type of settlement." "Indiscriminate" in this sense means there was no formal divisioning of the land before it was allocated to pioneer settlers. F. J. Marschner pointed out in *Land Use and Its Patterns in the United States* (1959):

> Although systematic land division had been proposed for Carolina and Georgia, preemption of land in the southern colonies proceeded with no attempt at conventionalized allocation of sites.... Each settler selected his homesite, which was usually near a stream and far enough from his nearest neighbor so that he would not encroach on his neighbor's claim.... The land surveys were seldom verified in the field and conflicting claims to land through overlapping were not unusual.

In view of such conditions, is it any wonder that the postcolonial landscape had a ragged or unregulated character? After a check of

Georgia's colonial grants in 1839, the state's surveyor-general informed the legislature that "the twenty-four counties existing in 1796 contained actually 8,717,960 acres of land, whereas the maps and records in the Surveyor-General's office show that in these counties there had been granted 29,097,866 acres."

The twenty-four eastern Georgia counties in existence in 1796 plus two that came into being in 1801 form the area known as the Headright Region. The term is drawn from the legal jargon of the day that granted acreages according to an individual's "headright," or right based on the number of people who would live and work on the land. Generally speaking, the Oconee River marked the western extent of this region of Georgia. The units granted and surveyed here vary widely in area, from family farms to those of several thousand acres. While rectilinearity is not entirely absent from the region's landscape compages, the headright scheme reflects irregularity rather than regularity in size and shape of land holdings.

PERFECTLY SQUARE FARMS

In the opening years of the nineteenth century, Georgians embarked on a revolutionary new program in granting land. Instead of allowing pioneers to take up individually surveyed headright parcels on newly gained Indian cessions, the state imposed a high degree of order on the settlement process. Beginning in 1803, the Indian-land cessions were each systematically surveyed into square, farm-sized lots by the state prior to their delivery to private owners in a series of free public-land lotteries. As a consequence, the western two-thirds of Georgia presents a remarkably rectilinear landscape to the airborne traveler. This area is known as the Land Lottery Region in recognition of the state land-lot surveyor's enduring imprint. It is an imprint of rectangularity not unlike that found in the Middle West and West of the United States.

Georgia is unique among the states in that a lottery system was employed to deliver about two-thirds of its public lands to private ownership without charge (federal lands were sold). Six land lotteries were held to dispose of the land in the western two-thirds of Georgia between 1805 and 1832. After a lottery was authorized by a legislative act, residents of the state could qualify by registering in the county where they lived. Names of those qualified were sent to the capital and placed in a lottery drum. The land to be granted had been surveyed and divided into land districts and each district was subdivided into perfectly square land lots of equal size. Tickets with land-district and land-lot numbers were placed in a second drum. Land commissioners drew a name and a land-lot ticket simultaneously from the two drums. The lucky winner then took out a grant to the lot drawn and, in due course, the property became his or hers. If the person selected did not want the land for some reason, it reverted to the state and was later sold at public auction—as were the irregularly shaped lots, called fractions. The public lotteries were not consistent with regard to the area of the lots dispensed or the orientation of the land-survey lines, but they gave western Georgia a landscape dominated by square land lots and marked rectangularity in other elements of the rural landscape.

PERFECTLY ROUND TOWNS

The two most potent and perfect forms in humankind's symbolic lexicon are the circle and square. These are precisely the forms chosen by Georgia's post-Revolutionary citizens to organize crucial aspects of their landscapes and lives. Square farms and fields, straight furrows and fence lines, straight lanes and roads with right-angle turnings dominate Georgia's landscapes in the vast rural space west of the Oconee and Altamaha rivers.

Over the whole of the state's 59,000 square miles and adjoining South Carolina a less immediately visible pattern of circular forms exists. These circles directly affect the lives of the Plantation South's urban citizens. More than 675 incorporated towns and cities in Georgia and South Carolina are currently, or have once been, perfectly circular by the terms of their limits of incorporation. Geographer Howard Schretter provided a map showing the distribution of "Circularly Bounded Incorporations" in nine southern states from Maryland to Texas, as they existed in 1960. Fully 503 of the 620 round towns that Schretter mapped were in Georgia and South Carolina. Of those, 379 or 75 percent were located in Georgia. A more recent survey of Georgia legislative acts revealed that 550 round towns were chartered in the period from 1810 to 1974.

Research indicates that the first round town in the United States was created when the east-central Georgia community of Warrenton received its charter in 1810. The Act to Incorporate Warrenton, the county seat of Warren County (founded Dec. 19, 1793), stated that the area of incorporation "shall extend to and take all the town lots that have been originally laid off . . . and also shall comprehend all the land within 300 yards of the said Court House it being the centre of the said corporation." Thus a circle with a radius of 300 yards centered on the Warren County courthouse described the area within which the government of Warrenton had the authority to tax and regulate.

Notice that there was no need to recite a long list of metes and bounds to describe the town's boundaries. Any place within 300 yards of the courthouse was in the town and any place farther away was out of town and not covered by town ordinances and regulations. It did not take long for other communities to follow Warrenton's lead. In 1811 Hartford and Marion, county seats of Pulaski and Twiggs counties, respectively, received charters describing circular corporate limits. In Hartford the radius was 300 yards, but in Marion it was 400 yards. In 1812 Sandersville in Washington County was incorporated with a courthouse-centered 400-yard circle bounding its limits.

Unsettled conditions caused by the War of 1812 put a temporary hold on town creation and not until 1815 did the legislature issue another new charter. This charter did not, however, create a new town. Rather, it was issued for the already existing home of the University of Georgia, Athens. Athens had been chartered in 1807; but no boundaries or limits to its jurisdiction had been specified at that time. In 1815 the Athens charter described a partial circle, with a one-mile radius centered at the chapel on the college campus. Rather than embracing all the area within one mile of the chapel, the boundary described excluded the large acreage located to the east of the North Oconee River; and not until 1840 was Athens's charter altered to embrace the total mile-radius area.

By 1861 some ninety-three Georgia towns adopted or were originally chartered with circular corporate limits. Of that number forty-eight of the circular limits centered on county courthouses. Town centers, crossings of designated streets or roads, public squares, designated residences, Masonic lodges, storehouses, designated trees, railroad depots and rail crossings, factory buildings, churches and schools, as well as land-lot corners and town blocks, were all employed as center points for circular corporate limits with radii that ranged from 300 yards (2) to one and a half miles (1), with most being half-mile circles (32).

As the traveler follows the described transect across the Plantation South most of the towns visited will be round towns. In many cases, such as Atlanta and Athens, piecemeal annexations destroyed their circularity. Where towns are no longer circular, arcuate remnants incorporated in their boundaries often bear witness to their earlier, more perfect shapes.

The traveler hoping to encounter streets radiating from town centers or circular boulevards like *ringstrasse* will be disappointed. In the round towns of the Plantation South circularity existed in the articulation of corporate power and influence over defined territory only. It did not extend to the layout of town plans and built environments within those circles. By and large round towns are dominated by gridiron street plans and square or nearly-square business and residential blocks.

What may possibly have begun as a symbolic statement bespeaking ideas of the common man's equality under law proved to be an immensely practical and very inexpensive way of bringing towns and cities into being. First and foremost, a survey was not required to determine and demarcate town boundaries on the ground—a considerable saving. A similar economy was realized in the legal verbiage required in writing charter statutes and other required descriptive documents.

The economy of language can be appreciated by comparing 1872 act to amend the charter of Athens with a similar act from 1918 which took the town's limits beyond a perfect circle:

1872—". . . the authority and jurisdiction of said city shall extend for a distance of two miles in an air-line in every direction from the college chapel;"
1918—". . . the corporate limits of the city of Athens . . . shall be extended so as to include on the northwestern limits of the present boundary . . . the following described additional territory. Said limits shall be extended as follows: Beginning at a point on the spur track of the Athens Railway and Electric Company. . . . ."

And the document continues for 224 words describing railway tracks and right-of-ways, center lines of streets, and old city boundaries.

During the decade of the 1830s the round-town phenomenon spread beyond the confines of its east-central Georgia hearth. By 1837, Whitesville and Corinth in Georgia's westernmost counties were chartered as round towns. In the neighboring states Alabama, South Carolina, and North Carolina, round towns had also been created. In South Carolina the round towns of Lancaster, Edgefield, Orangeburg, Greenville, Abbeville, and Anderson served as the county seats of the up-country counties bearing their names. By the decade of the 1960s, when geographer Howard Schretter completed his investigation, he found the circular corporate limit to be "a phenomenon of the Southeastern United States with all but twelve of the total places bounded by circles located in the seaboard states from Maryland to Alabama."

Is the round town a form of civil organization that, because of its simplicity and economy, could be employed to good effect in developing areas of today's world? It has certainly stood the test of time and change in the Plantation South.

PART TWO

The Itinerary

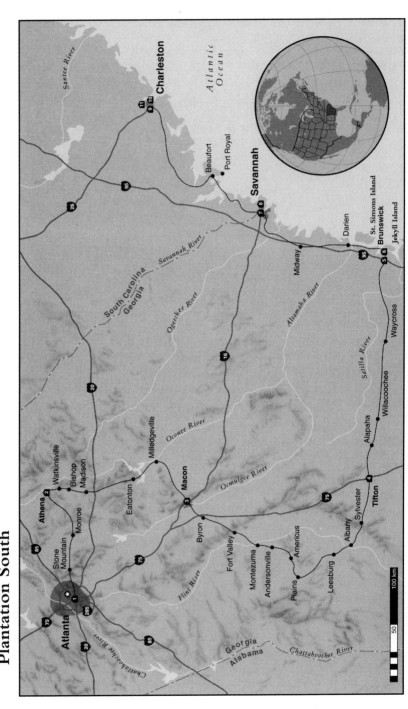

⚠ Highlights of the Region

This trip begins in Atlanta, a northernized metropolis with little evidence of its Old South origins. Local attractions include the Martin Luther King Center, shrine of civil-rights activism of the 1960s; the Cyclorama, a massive painting of the Battle of Atlanta which evokes the strategic importance of Atlanta in the short-lived Confederate States of America; and the Carter Center, headquarters of the many international activities of former president Jimmy Carter. The route to Athens passes the great granite mass of Stone Mountain, emblazoned with heroes of the Confederacy.

From Athens, the route to Macon goes through the Georgia Piedmont via Bostwick, an old farming hamlet, and Madison and Eatonton, whose historic districts boast fine antebellum homes. Eatonton was the hometown of two famous Georgia writers: Joel Chandler Harris, of B'rer Rabbit fame, and Alice Walker, author of *The Color Purple*. A quick tour of Milledgeville, capital of antebellum Georgia, includes the Gothic Revival capitol building and Federal-style houses.

The city of Macon is in the sand hills at the fall line. In addition to a city tour, visits are suggested to the Ocmulgee Mounds, from the precontact Mississippian period; nearby Clinton, once famous as a cotton-gin manufacturing center; and Jarrell Plantation, preserved as a museum with a collection of farm buildings from the antebellum period to modern times.

The trip from Macon to Tifton goes through the upper Coastal Plain via Andersonville, a Civil War prison and national cemetery; Plains, home of President Jimmy Carter; and into Georgia's modern farming belt. In late summer, cotton-picking machines and cotton gins are in operation. Tifton has an agricultural experiment station and is the home of Agrirama, a living agricultural museum.

THE PLANTATION SOUTH

Leaving the Tifton Uplands, the route to Jekyll Island crosses the lower Coastal Plain, often referred to as the "pine barrens," an area of thin soil and little agriculture. A diversion can be made into the northern edge of Okeefenokee Swamp, a vast wetland noted for its special ecology and wildlife.

Jekyll Island is a good place to examine the barrier islands, and see traces of historic rice plantations, some now set aside as waterfowl refuges. At St. Simons visit the remains of Britain's Fort Frederica, the site of the Battle of Bloody Marsh, the final Spanish effort to oust the British from Georgia. Darien, on the north fork of the Altamaha, was settled by Scots, and here stands a replica of Fort George.

From Jekyll Island, continue north to Savannah, stopping enroute to visit the Midway Congregational Church (now Presbyterian), established by settlers of New England ancestry; Hardwick, on the Ogeechee River, once proposed as an alternate site to Savannah; and Fort McAllister, an earthen Civil War fort.

Savannah, the oldest settlement in Georgia, was the cultural and economic center throughout the antebellum period. However, it lost political importance after the Revolution, and the capital was shifted north. Savannah's gridiron pattern with alternating squares provides a distinctive urban plan. Its rowhouses provide an ambience suggestive of London. Nearby Ebenezer was a Salzburger settlement, once a center for religious refugees.

The route to Charleston is via the low country and Beaufort–Port Royal, an area rich in French and Spanish history as well as early British settlement. On St. Helena Island, center of Gullah culture, the Penn Center attempts to maintain African-American cultural traditions.

The traverse terminates in Charleston, South Carolina, cultural and spiritual hearth of the Plantation South. Already in economic and political decline, the Civil War brought economic disaster and poverty. In the absence of economic development, Charlestonians have clung to their rich heritage of public and private buildings, which today forms the backbone of a thriving tourist industry. Eighteenth-century plantations lie along the Ashley River as does Charles Towne Landing, where a permanent British settlement was made in 1670.

The plantation economy which made possible the visible signs of wealth in this region has disappeared, but it left a legacy of public and private buildings, rich in evocative memory and associations. These belong to blacks as well as whites; from the work of both were they created.

In terms of topographic areas of Georgia and South Carolina, this tour spends two days in the Georgia Piedmont, one day at the fall line and in the sand hills, one day in the upper Coastal Plain, one day in the lower Coastal Plain, and six days on barrier islands and in the low country. The Appalachian Mountain area of Georgia begins about fifty miles north of Athens, and is not included in this traverse. In the antebellum period, it was mainly an area of small subsistence farms, as was the lower Coastal Plain.

△ *Day One*

ATLANTA, GEORGIA

Atlanta, the capital of the state of Georgia and seat of Fulton County, is the largest city in the state. Forty-four percent of the state population lives in the metropolitan area. Not only a major regional center in government, finance, business, education, medicine, air and ground transportation, and distribution, Atlanta is also headquarters for a number of national corporations.

In the process of growth, Atlanta built over its antebellum origins, so that its skyline, as well as its spirit, has more in common with other large cities in other states than with those in Georgia. Since World War II extensive suburban growth has made it harder to maintain a distinctive urban core. Skyscrapers and office parks mushroom far from the traditional central business district. White flight to the suburbs has created a black majority in the city of Atlanta and in Fulton County. Both governments now have black chief executives, Mayor Maynard Jackson and County Commissioner Michael Lomax, dedicated to the continuing business growth of the Atlanta area.

History

The first mention of the area that became Atlanta dates from 1782, when the up-country was still troubled by Loyalist-Patriot clashes and Indian attacks were a constant concern. In 1813, a fort was ordered built here at a site between the Chattahoochee and Nancy's

Creek near a Creek village known as the Standing Peachtree. Until the Cherokee removal in the winter of 1838–39 by U.S. soldiers, the Standing Peachtree was an important trading post and entry point to the Cherokee Nation to the northwest. Although Indian land south of the Chattahoochee had been ceded in 1821, white settlement was surprisingly slow. The first white settler to build a permanent dwelling in the area was Hardy Ivy in 1833.

The growth of the area came about as a result of railroads pushing into the interior to cotton lands in the Piedmont. The economic feasibility of hauling cotton by rail had already been demonstrated by the time Georgia chartered its first railroad, the Georgia Railroad, in 1833; and the rails spread rapidly. By 1860, except the northeast mountain area and the southeast pine barrens, all of Georgia had rail service; and Atlanta, incorporated in 1847, had become a major railroad terminus. As such, it was destined to become a major supply depot and military production center.

Thus Atlanta was a logical target for Union attack after the fall of Nashville in 1862, but the war was diverted west to the Mississippi. It was not until May 5, 1864, that a 100,000-man Union army under the command of General William Tecumseh Sherman moved south from Chattanooga to begin the campaign to take Atlanta. He was opposed by Confederate General Joseph E. Johnston with a force of 60,000. Although Johnston fought a classic delaying action, he was constantly pushed back by superior force. Impatient with his tactics, Confederate President Jefferson Davis relieved Johnston of command on July 18th and replaced him with General John B. Hood, who had a reputation as an impetuous fighter.

Hood attacked Sherman a number of times, including the so-called Battle of Atlanta on July 22; but he was unable to prevent Sherman from besieging the heavily fortified city. On August 25, Sherman moved his main force west and south to cut the railroad at Jonesboro. A Confederate counterattack there was unsuccessful; and, after destroying vast stores of military supplies, Hood evacuated Atlanta. Sherman entered the city on September 2nd.

General Hood remained in the vicinity for about a month, and then moved northwest toward Nashville, Tennessee. This left no Confederate army in the field in Georgia or South Carolina; and on

November 15, 1864, Sherman, with a force of 68,000, began his famous, virtually unopposed March to the Sea. Before departing Atlanta, everything of a public nature that could be of aid to the Confederacy was destroyed. Only 400 of the 3,600 houses within the city limits were left standing. The City Hall and Council House and a few of the churches remained; but almost all the stores, banks, hotels, office buildings, factories, foundries, and railroad facilities had been destroyed.

Despite the massive destruction, revival got underway in Atlanta very quickly after the end of the war. First of all, the railroads were soon operating; and two other factors helped: Atlanta became the seat of the military governor and the state capital. Under the 1867 plans for Military Reconstruction Atlanta was made the seat of a military district consisting of South Carolina, Georgia, and Florida, which meant that federal soldiers and employees in Atlanta had hard U.S. currency, scarcely available elsewhere. Also, Governor John Pope ordered the state legislature to meet in Atlanta when he learned that the innkeepers of Milledgeville, the state capital, had allegedly refused to accommodate black elected representatives. The Republican Convention of 1868 made Atlanta the state capital and this was reaffirmed by the Constitution of 1877, enacted after the formal end of Reconstruction in the South. By 1880, Atlanta had a population of over 37,000, and had become larger than Savannah.

Of the many interesting places to visit in Atlanta, we have chosen to highlight four: the Martin Luther King, Jr., Center, the Cyclorama, the State Capitol, and the Jimmy Carter Museum and Library.

Martin Luther King, Jr., Center for Nonviolent Social Change

The center, at 449 Auburn Avenue, is located in a two-block area listed in the National Register of Historic Places. The red-brick complex was dedicated in 1982 and contains the tomb of Martin Luther King, Jr. A brochure outlining a self-guided tour is avail-

Atlanta, Georgia

1. Ebenezer Baptist Church
2. Martin Luther King, Jr. Center
3. Cyclorama
4. Carter Presidential Center
5. City Hall
6. State Capitol
7. CNN Center

32   THE PLANTATION SOUTH

Martin Luther King, Jr., spent the first twelve years of his life in this house at 501 Auburn Street in Atlanta. King's grandfather bought the house in 1909, and the family occupied the house for three generations. King grew up in a neighborhood of recognized leaders of Atlanta's black community.

able from the receptionist. Nearby is the home where Dr. King was born on January 15, 1929, and the Ebenezer Baptist Church, associated with his father and grandfather and an unofficial center for Civil Rights activities during the 1960s. A donation is suggested for admission to the King Center and for entrance to the church. There is no charge to visit the birthplace, which is managed by the National Park Service.

King, Jr., was educated in the Atlanta public schools. He then entered Morehouse College, and was ordained by the National Baptist Church before he graduated. With a bachelor's degree,

King went on to Crozer Theological Seminary and then to Boston University, where he was awarded a doctorate in theology in 1955. In Boston, he met Coretta Scott, whom he married on June 18, 1953.

In 1955 Dr. King was head of the Montgomery Improvement Association, which was leading the bus boycott by blacks against segregated seating. (The Rosa Parks Room in Freedom Hall at the King Center honors Mrs. Parks, whose refusal to relinquish her seat to a white passenger on orders of the bus driver led to the boycott.) During this period, Dr. King perfected the rhetoric for which he became famous and his policies of nonviolence patterned on the principles of Mohandas K. Gandhi of India.

In 1957, Dr. King helped establish the Southern Christian Leadership Conference and until his assassination in 1968, he was the acknowledged spokesman for the Civil Rights movement. King delivered the famous "I Have a Dream" speech in Washington in 1963; in 1964, he was awarded the Nobel Peace Prize.

In Memphis to lead demonstrations in support of black sanitation workers, Dr. King was assassinated on April 4, 1968. His remains were transferred to the present site in 1970.

Martin Luther King Day, a national holiday, is observed on the third Monday in January.

The Cyclorama

An impressive panoramic painting of the Battle of Atlanta—July 22, 1864—is on display in Grant Park, in southeast Atlanta. The building that houses it also contains a Civil War museum, whose gift shop has a fine selection of books dealing with the Civil War. A motion picture depicting the campaign that led up to the Battle of Atlanta is shown every half hour. There is an admission fee.

In the nineteenth century, the mass production of vast circular paintings developed into a practical business enterprise. The Atlanta Cyclorama, painted by the American Panorama Studio of Milwaukee, is 400 feet in circumference and 50 feet tall, and weighs 1,800 pounds. One of several cycloramas of Civil War battles, it was commissioned in 1883 as a campaign document by General John A. Logan to support his candidacy for vice-president in 1884. The battle is depicted at a crucial moment—4:30 p.m.—when General Logan's Union forces under Major General Francis P. Blair counterattacked and broke the Confederate advance. The prominence given to Logan in the painting reflects the political purposes for which the cyclorama was painted.

In the early 1890s the Battle of Atlanta cyclorama was brought to Atlanta, and remained on exhibition until 1898. G. V. Gress, an Atlanta lumber merchant, purchased the painting and presented it to the city, which subsequently built the Cyclorama building to house it. In 1937, with assistance of federal funds, professional painters and sculptors added a three-dimensional effect. This blending of the then-stationary central platform into the foreground of the painting was so realistic that even the keenest scrutiny did not reveal where the foreground figures gave way to the actual painting. Unfortunately, the technology that introduced the three-dimensional effect contributed to the deterioration of the painting.

During the second term of Mayor Maynard Jackson, a multimillion dollar renovation of the building, painting, and three-dimensional figures was undertaken. The massive painting was strengthened with additional backing, the interior viewing platform was replaced with a revolving seating platform, and a timed, automated show was developed to highlight aspects of the painting.

Grant Park lies in what had been the defensive perimeter of Atlanta and is covered with the remains of many breastworks. The crown of hill at the southeast corner of the park is the location of Fort Walker, named after William H. T. Walker, a Confederate general killed in the Battle of Atlanta.

The State Capitol

The capitol, on Washington Street between Hunter and Mitchell, is open to the public every day. Patterned after the national capitol, it was completed in 1889 at the cost of a million dollars, within the amount budgeted. To the regret of many modern Georgians who would have preferred Georgia granite and marble, the building is made of Indiana limestone.

The dome, rising 237 feet above the ground, is covered with gold from Dahlonega, Georgia, site of a gold rush in 1829—first in the country—which brought several thousand whites into what was still Cherokee territory. The discovery of gold in Georgia helped seal the fate of the Indians east of the Mississippi; Congress enacted the Indian Removal Act of 1830.

The capitol contains chambers for the Senate and House of Representatives and houses the governor's offices, a natural science display, the Hall of Flags, and the Hall of Fame honoring outstanding Georgians. Several state office buildings and the administrative offices for the city and the county are located nearby.

Jimmy Carter Library and Museum (The Carter Center)

The Carter Center at 1 Copenhill is open daily to the public. The complex consists of the presidential library, administered by the U.S. National Archives; a museum, which portrays the election and presidency of Jimmy Carter in pictures, video, film, and displays; and offices from which the center carries on its many international programs. The west garden gives a dramatic view of the Atlanta skyline. Facilities include a cafeteria and a gift shop. An admission fee is charged.

The construction of the Carter Center facilities was funded by $25 million in donations from individuals, foundations, and corporations. Dedicated on October 1, 1986, the complex has four interconnected buildings set in thirty acres. The Carter Library and Museum is deeded to the federal government.

The center is also home to a consortium of nonprofit organizations that seek to reduce conflict, mitigate suffering, and promote better health and agricultural services in developing countries. The Task Force for Child Survival was formed in 1984 to facilitate immunization and other efforts in developing countries. The Carter-Menil Human Rights Foundation supports and supplements the work of the nongovernmental agencies, and each December awards a $100,000 prize to an individual or group that exemplifies human-rights leadership. The Carter Center of Emory University (CCEU) addresses issues of public policy through nonpartisan study and research. An unusual array of leading scholars, diplomats, and policy makers have contributed to the work of the CCEU since 1982. Study programs cover such topics as international and domestic health, human rights, and conflict resolution in many parts of the world.

Other Attractions

There are many things to do in Atlanta—a visitor with the time could spend several days sightseeing. The Atlanta History Center in Buckhead manages the McElreath Hall Museum, Swann House, and the Tullie Smith House. The Wren's Nest, home of Joel Chandler Harris, creator of the stories of Uncle Remus and Br'er Rabbit, is open to the public. Tours are available at the CNN Center, home of the Cable News Network and Headline News. *Atlanta Now,* the official visitors guide of the Atlanta Convention & Visitors Bureau, includes a calendar of events, maps, city and area attractions, sightseeing tours, dining suggestions, and other useful information.

Atlanta, Georgia, to Athens, Georgia

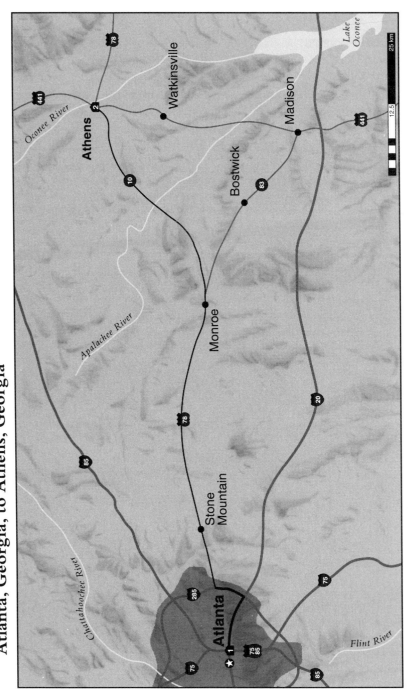

△ Day Two

ATLANTA TO ATHENS, GEORGIA
(75 miles, U.S. 78)

Stone Mountain

Twenty miles east of Atlanta is Stone Mountain, a geological landmark, amusement park, and Confederate memorial. The granite dome rises 650 feet above the surrounding Piedmont Plateau, and is about 2 miles long with a base circumference of 7 miles. The mass which erosion has exposed is only a fraction of a vast magma pluton which was formed about 300 million years ago. Sometimes this area is called a batholith, but Dr. Gilles Allard of the University of Georgia Department of Geology prefers to describe it as a stock, since the granitic area, which merges into an extensive field of gneiss toward Monroe and Athens, is less than 40 miles square. The dome is almost bare; its shape is elliptical. The north face, in which the Confederate Memorial is carved, is sheer, with pronounced streaking from iron oxide carried down by rain water. The slope to the top is gentle from the west, and may be gained by a walking trail.

HISTORY

Archaeological evidence indicates Indian use of Stone Mountain as early as the Woodland period. Stone Mountain and the Atlanta area were ceded by the Creeks in 1821. By 1825, white settlement supported a stagecoach stop and inn at the mountain's western base. The place soon attracted visitors; getting a grand view of the countryside then, as now, was a favorite pastime. In 1842 Cloud's Tower, 165 feet high, was erected on the summit. In 1880, the mountain became the property of Samuel Hoyt Venable, who quarried fine quality granite as well as coarser stone for road construction. In 1916 the north slope was deeded to the Stone Mountain Memorial Association for the carving of a great Confederate monument. In 1958, the state of Georgia acquired the property, and the surrounding area was converted into a recreation area. Among popular amusements are a summit ride by cable car and a circuit of the base by train.

THE CONFEDERATE MEMORIAL

A memorial at Stone Mountain was originally a project of the United Daughters of the Confederacy, which had been organized in 1894. None kept the idea of a Southern past more alive than these women of the former Confederacy.

The original conception of the memorial was far greater than the resources available. Across the north face, in a panorama 1,350 feet long, Confederate forces would pass in review; at the base would be a memorial hall and a sarcophagus honoring all Confederate dead.

American entry into World War I interrupted the project, and not until 1923 did sculptor Gutzon Borglum begin the carving. By 1924, the head and shoulders of the central figure, General Robert E. Lee, had begun to emerge and in celebration a breakfast was held on his left shoulder for fifty guests. Soon after, in a dispute with the memorial association, Borglum's contract was canceled; in a rage, he destroyed his models and left Georgia. [The gigantic heads of presidents at Mount Rushmore, South Dakota, are a later Borglum.]

In 1925, Augustus Lukeman was engaged; and in 1928 he submitted a new, more modest plan. The old work was blasted away and new central figures of Davis, Lee, and Jackson were begun. But with the onset of the Depression, money ran out and the work was abandoned.

After the state-financed Stone Mountain Commission acquired the Venable property in 1958, plans were revived to complete the partially carved Lukeman figures; Walter Hancock of Gloucester, Massachusetts, was engaged as sculptor. Actual carving began in 1968, and the monument was dedicated on May 9, 1970. The vast carving, dwarfed by the mass of the mountain, reminds the history-minded of the frailty of human dreams.

In the Stone Mountain Park is an impressive display of houses, which constitutes an antebellum architectural and interior design museum. A number of old houses in a wide variety of styles have been moved here and make this a worthwhile stop.

In 1996 Stone Mountain will be the venue of some of the Atlanta-based Olympic Games.

Stone Mountain to Monroe, Georgia, 28 miles, U.S. 78

Monroe, the county seat of Walton County, was named in 1821 in honor of President James Monroe. Walton County is named for George Walton, a delegate to the Continental Congress from Georgia in 1776, and a signer of the Articles of Confederation and of the Declaration of Independence.

Walton was created as Georgia's 46th county by an 1818 act of the legislature. The county was surveyed in 250-acre lots, resulting in 910 square lots and 1,108 irregular parcels. The lots were distributed as part of the 1820 Land Lottery.

The land in Walton County is rolling, with elevations rising from 750 feet in the southeast to over 1,000 feet in the northwest. Three erosional remnants rise from 200 to 300 feet above the peneplain: Turkey Creek Mountain, Jack's Creek Mountain, and Alcovy.

The Apalachee, which forms the eastern boundary of Walton, was the eastern boundary of the Creek Nation established by the Treaty of New York in 1790. White settlers in the area constantly violated the boundary and pressed westward, especially at the High Shoals crossing. This led to an Indian trail known as Rogue Road. The land which became Walton County was part of the 1802 Creek Cession.

Into the 1960s, Walton remained a major cotton-producing county and the town of Monroe had a number of large cotton mills. The county has since lost its rural characteristics, and the city most of its mills; today it is a bedroom community to the Atlanta metropolitan area. Running north-south along the high ridge of Rogue Road, Ga 11 passes in front of the handsome county courthouse with the Confederate monument out front. Many of the older buildings of the town are along this road, as are the remnants of the old business and shopping area.

Athens, home of the University of Georgia, is 27 miles east of Monroe This stretch of the highway is called the Moina Michael Highway, after a Walton native born near Good Hope in 1869. During World War I, inspired by John McCrae's poem "In Flanders Field," she scribbled a poem "We Shall Keep the Faith." Out of this grew the idea of the Flanders Field poppy, which became the memorial emblem for veterans.

Day Three

ATHENS TO MACON, GEORGIA (115 miles)

Athens

Writing about the University of Georgia, historian E. Merton Coulter stressed that "Those in authority who made the University also made the Town." The traveler quickly realizes that the modern relationship between Athens and the University of Georgia is no less strong than it was in 1801 when John Milledge purchased 633 hilly acres on the west side of the North Oconee River overlooking a water-driven mill complex. In the account announcing the transaction to the citizens of Georgia, the *Augusta Chronicle* for July 25, 1801, reported, "The Senatus Academicus, having designated the county in which the University should be established, named a committee to select the site and contract for a building."

After repairing to the frontier county of Jackson, the committee examined "a number of situations as well upon the tracts belonging to the University." When their examinations and deliberations were concluded "the vote was unanimous in favour of a place belonging to Mr. Daniel Easley, at the Cedar Shoals." The tract "was called Athens" and "the square of the University containing thirty-six acres and a half is laid off so as to comprehend the site, the houses, the orchards and the spring, together with a due proportion of the woodland."

Milledge and the other committee members selected a site where the seeds of a nascent community were already planted, rather than in the midst of the still-to-be-tamed wilderness where the university owned land. In a short time lots were laid out for a town on the north side of the university campus and Athens began to form.

CITY TOUR

As the visitor leaves the campus heading west he or she should turn north onto Lumpkin Street and travel along the edge of the historic campus. A series of high-rise student dormitories on the left gives clear evidence of the campus sprawl that continues to impact Athens.

On the right, at the corner of Cedar Street, is an attractive roofed shelter that once served as a streetcar stop in another part of Athens. Lumpkin Street presents a scene generally known as "campus fringe." Large academic buildings dominate the right-hand view while sorority and fraternity houses and denominational student centers front on the left. The landscaped complex midway up the hill on the right is the home of the Garden Club of Georgia, the first garden club in the United States.

Turn right onto East Broad Street where Athens's central business district faces the historic North Campus. "Town" is on the left and "gown" on the right of this busy thoroughfare. The large anchor marks Athens as the home of the U.S. Navy Supply Corps officer training facility. Visitors with sufficient time should stop at the museum located on the campus of the *Navy Supply Corps School* on Prince Avenue.

Turn right on Spring Street, named for the "copious spring of excellent water" that helped make the site of Athens so attractive to John Milledge and his associates. The spring continues to flow copiously beneath the paving of this commercial area of modern Athens. It joins the Oconee River just upstream of the *Cedar Shoals mill site*. To reach the site, continue across the railroad tracks. Here, O'Malley's Tavern, a popular student watering place, is an interesting example of the adaptive use of an old industrial building. Daniel Easley was operating a mill complex in 1801 when the adjacent acres were purchased for the university and townsite. In 1833 the Athens Cotton and Wool Factory was constructed at the mill site. This textile mill described by the famous

44 THE PLANTATION SOUTH

English traveler James Silk Buckingham, who visited Athens in 1839:

> On the banks of the Oconee river . . . are three cotton factories all worked by water power and used for spinning yarn, and weaving cloth of coarse qualities for local consumption only. I visited one of these, and ascertained that the other two were very similar to it in size and operations. In each of them there are employed from 80 to 100 persons, and about an equal number of white and black. In one of them, the blacks are the property of the millowner, but in the other two they are the slaves of planters, hired out at monthly wages to work in the factory. There is no difficulty among them on account of colour, the white girls working in the same room and at the same loom with the black girls, and boys of each colour, as well as men and women, working together without apparent repugnance or objection. (*The Slave States of America,* London, 1842, p. 112)

The mill was operated successfully through the Civil War and in 1890 it was reported to have 10,000 spindles; it finally ceased production in the 1920s and the machinery was sold off. In the late 1970s the building was converted to a minimall and restaurant. The retail shops failed but the tavern and a health club seem to be flourishing in symbiosis.

Proceed under the railroad on Baldwin Street and turn left on East Campus Road. At the intersection with Milledge Avenue turn left to the *State Botanical Garden.* After making the Botanical Garden loop turn right and continue to the *White Mansion* in the town of Whitehall.

This imposing brick mansion was the home of local industrialist John R. White. It was built in the 1890s to replace the original homeplace of White's father who had arrived in Athens from County Antrim, Ireland, in the mid-1830s. The mansion and White Experimental Forest are now owned by the University of Georgia. The house was restored in the 1970s with proceeds from timber sales.

Across the railroad tracks are a number of small cottages that once housed the workers employed at the Georgia Factory textile mill, the area's first factory-scale enterprise. The mill was built on the North Oconee River, about four miles downstream from Athens. As the following news account from the March 31, 1829, edition of the *Athenian* newspaper reveals, many Southerners were less than enthusiastic when it came to industrial development in their heretofore agrarian region. Keep in mind this extract was written over thirty years *before* the outbreak of the Civil War:

> On Thursday last, at 12 o'clock, we understand the ceremony of removing the first earth for the foundation of a COTTON FACTORY, was performed by Judge Clayton, at Mr. Thomas Moore's Mills, four miles below this place. We notice this transaction not so much for what might seem an idle parade, but for that which we conceive to be the first step towards a very important change in the productive industry of the country. The agricultural character of the south has been compelled, by a very unwise policy of the General Government, to partake of that manufacturing spirit which seems to have been forced down upon the nation at the expense of every other interest. A sense of safety and a feeling of independence, combined, doubtless, with an expectation of profit, have urged gentlemen to an undertaking, against which their political convictions are most unquestionably at war. And we are authorised to state, that those sentiments have, by no means, undergone a change; that their project is certainly not to give countenance to a system which they have always denounced; but it is to be regarded as a measure unquestionably defensive. The experiment is an eventful one—the period, however, has arrived when no other alternative is left but to strike for commercial freedom. May their effort be attended with success....

Athens, Georgia, to Macon, Georgia

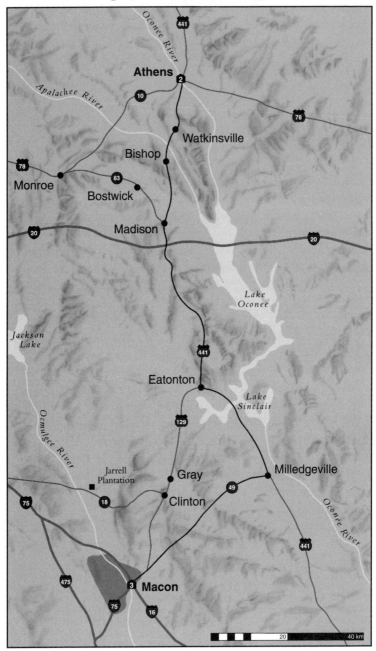

Athens to Madison, U.S. 441/Ga. 83, 41 miles

Upon exiting the White Mansion turn left and follow Simonton Bridge Road to Watkinsville. When Clarke County was created in 1801 it included the area now embraced by Oconee County and the county seat was Watkinsville. In 1871, the county government was transferred to Athens by a legislative act. The citizens in the southern part of the county were furious and responded by lobbying for the creation of a separate county. Oconee County was established in 1875 and Watkinsville resumed its position as a center of civic and business life. The county was named for the river which forms the county's eastern boundary. A group of the Oconee tribe living along the Oconee River in the seventeenth century probably lent its name to the stream.

Turn right onto US 441 and proceed to the historic *Eagle Tavern Museum*. Informative exhibits are housed in this restored late eighteenth-century building; and restrooms are located in an authentic log cabin just behind. The tavern had a long history as a stagecoach stop and hotel and, by 1857, periodic additions had added sixteen rooms to its four-room plain-style nucleus. The Georgia Historical Commission accepted the rambling twenty-room hotel in 1956. After careful study a restoration of the oldest portion was undertaken and the later, rather haphazard, additions were eliminated. Eagle Tavern is now maintained as the Oconee County branch of the State Welcome Center.

Departing the Eagle Tavern the traveler should turn left and go south on the main street, US 441. Opposite the tavern is the Oconee County Court House, a fine example of a small, FDR-era public building. Attractive peach orchards line both sides of the road. Consumers are offered lower prices if they will pick their own fruit. Although still known as "The Peach State," Georgia is regularly exceeded in peach production by South Carolina. Keep a sharp eye open for a vineyard on the left side of the highway. The grapes are of the muscadine variety and not used for wine.

Turn right at Bishop—a town typical of hundreds of small service and trade centers located along Georgia's railroads. Such towns are no longer able to compete for business with larger towns like Watkinsville and Madison. The local road leads to Bostwick. *Bostwick* was chosen for a visit because it typifies many of the changes that have taken place in small service-center towns across the Old Piedmont Plantation Belt. First incorporated in 1902, the town was named after its chief promoter and builder, John Bostwick. Among his many accomplishments was the building of the Susie Agnes Hotel, now the community center. Walk around Bostwick and visit the cotton gin and the general store, which once employed a staff of eighteen. Try to imagine what the town was like more than fifty years ago. The arrangement of Bostwick's remaining buildings indicate that the town was once served by rail.

Drive south from Bostwick on Ga. 83 to *Nolan's Store* and crossroad. Here the visitor is at the heart of what was once a 2,000-acre cotton plantation that operated from the opening decade of the nineteenth century through the 1970s. Several features bear witness to that long history. The Nolan family bought the plantation in 1856 and the property still remains in the Nolan family. Two main houses can be seen in close proximity to a badly dilapidated cotton gin. The original plain-style homeplace is a typical I-form house with two rooms over two and a broad central hall. Built early in the last century, it is rapidly falling into ruins through neglect. The newer, more impressive house was built about 1910 near the commissary and several tenant houses

Take the road to Apalachee and turn south on US. 441 to Madison. *Apalachee* is a version of a toponym that appeared on Mercator's world map of 1569 and is usually regarded as Georgia's oldest recorded Indian name. The town takes its name from the nearby Apalachee River which was sometimes termed the South Branch of the Oconee River.

Madison to Eatonton, U.S. 441, 19 miles

Madison, the county seat of Morgan County, was founded in 1809. The traveler should begin his or her examination of this historic town by visiting *Heritage Hall*. This Greek Revival-style home was built about 1842 for a prominent physician, Dr. Elijah Evans.

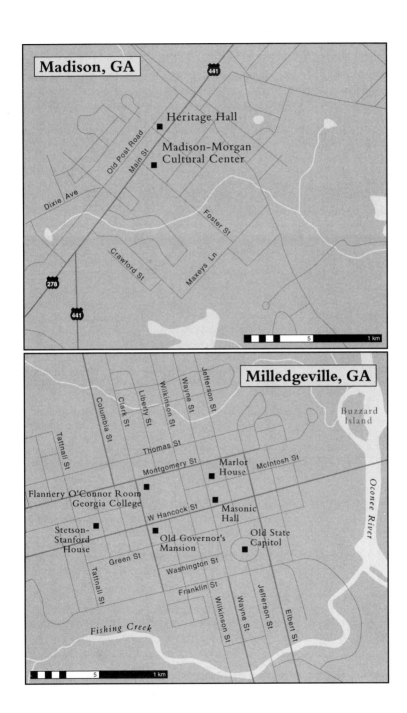

It was moved from an in-town "farm" lot to its present location in 1912. Heritage Hall served as a private residence from its construction until 1977, when it was presented to the Morgan County Historical Society.

When the geographer John Melish visited this area in 1809 he admitted that he "was really surprised to observe the number of settlements that had been made in the short space of four years . . . I was hardly ever out of sight of a plantation." Madison, he noted, "is now a thriving place, having a court-house, a number of dwelling houses, three taverns, and as many stores."

George White, in his *Statistics of the State of Georgia* published in 1849, was moved to term Madison "the most cultured and aristocratic town on the stagecoach route from Charleston to New Orleans." Miraculously, the town, with its impressive collection of Greek Revival- and Federal-style buildings, was spared by the Union army of General William T. Sherman in its infamous March to the Sea. According to local lore, Madison escaped devastation because Sherman responded to a plea from Senator Joshua Hill, a Southerner who had voted against secession. Whatever the reason, the modern visitor cannot help but mutter a silent thanks.

After walking around Madison's courthouse square the visitor should tour the spacious residential blocks on the short drive south to the *Madison-Morgan Cultural Center*. A center for the visual, decorative, and performing arts, the Cultural Center is housed in a magnificent 1895 Romanesque Revival-style building. Originally a school, the center's imposing brick structure was completely restored for its present role in 1976. Visit the historical museum, and galleries, and beautifully restored auditorium where a full schedule of plays, recitals, and lectures are presented.

Leave the Cultural Center, turn left, and follow US 441 south. Near the interchange of I-20 is much recent commercial development. It has provided economic opportunities to Madisonians without severely impacting the historic quality of the town center.

For the next several miles, there are a number of dairy and beef-cattle operations. Huge rolled hay bales are often stored in the open rather than in traditional barns. The sharp-eyed will notice small graveyards that remain from the era when rural population densities were far greater than those of today. Here on the Piedmont, as over most of Georgia, the rural scene is dominated by pine trees—the state's largest crop. An occasional roadside sign will announce "Trees Grow Jobs." This in what was once of the heart of the Plantation South.

Just over the boundary in Putnam County, turn right into the *Rock Eagle 4-H Center*. The center is operated by the Cooperative Extension Service of the University of Georgia. Each year tens of thousands of Georgia's younger citizens gather here for a variety of educational and recreational experiences.

The center takes its name from the internationally known bird-shaped effigy mound. The visitor should climb the nearby tower and view the mound of quartz rocks from above. There is some debate as to whether the ancient builders of the effigy intended it to represent an eagle or a turkey vulture. Archaeologists feel that the effigy dates from the Woodland period (1000 B.C.–A.D. 900). A similar effigy a few miles away is not accessible. The wingspread of the Rock Eagle effigy is 120 feet while the other is some 12 feet larger. Indians inhabiting Georgia at the time of earliest contact by Europeans evidenced no knowledge of these effigies or their builders.

Depart Rock Eagle Center and turn south on US 441 to *Eatonton*, the county seat of Putnam County. The town and county are named for Revolutionary War heros: General George Israel Putnam and General William Eaton. The county was founded in 1807 and city incorporated in 1809. Dominating the center of Eatonton is the handsome bulk of the Putnam County Courthouse set in the center of its square. Built in 1905 to replace an earlier building, the courthouse was designed to serve as a grand "county capitol" in a neoclassical style. Four entrances, of equal importance, face in four directions. Notice the sculpture of Br'er Rabbit on the east lawn of the square. This famous character was created by Joel Chandler Harris, one of the first authors to record black folk literature. Harris (1848–1908) was born in Putnam County.

Heading east from the square, Sumter Street becomes Ga 16, here called Sparta Road. About 7 miles east turn left on Old Phoenix Road. Continue north to *Turnwold Plantation* house. In the antebellum period, Turnwold formed a plantation settlement that included a hat factory, a tannery, a distillery, and a printing press, in addition to the usual farming activities. As a boy Joel Chandler Harris lived at Turnwold where he served as an apprentice printer under J. A. Turner, editor and publisher of the *Countryman*. Harris's book *On the Plantation* deals with this formative period in his career.

Continue north on Old Phoenix Road when departing Turnwold. At Ward Chapel Road turn right to Ward Chapel's African Methodist Church. This is the area that inspired Putnam County's other world-renowned author, Alice Walker. Probably best known for her Pulitzer Prize-winning novel *The Color Purple*, Alice Walker grew up in this neighborhood. Regrettably the house where she was born in 1944 was razed. In the small cemetery across the road from the church, however, are the graves of her father and his parents. A "Color Purple" Heritage Trail is under development.

Eatonton to Milledgeville, U.S. 441, 20 miles

Return to US 441 and travel south. The route to Milledgeville is sparsely populated, and there is little farming. The major activity is lumbering. Stop at the Welcome Center gazebo about 2 miles beyond Flat Rock, at Little River State Park and Lake Sinclair. Brochures on Milledgeville and the area are available here at all times. Especially helpful is "A Guide to Milledgeville: Georgia's Antebellum Capital" with thirty illustrations.

Lake Sinclair, on the Oconee River, was formed by the Sinclair Dam, completed by the Georgia Power Company in 1953. The lake covers 15,000 acres, has 500 miles of shoreline, and serves as a location for lakeside homes and water recreation. The annual September Ski Show is a popular event.

Twenty miles southeast of Eatonton on US 441 lies Milledgeville, seat of Baldwin County and capital of Georgia from 1803 to 1868. The city was named after John Milledge, governor of Georgia from 1802 to 1806. The original sixteen blocks, laid out in 1803, boast one of the highest concentrations of Federal-period houses in the United States. The National Trust for Historic Preservation has named Milledgeville the only surviving example of a Federal city. Most houses in the historic area are two-story, clapboard houses: two rooms over two or four over four, with a stair in the central hall. Most have a balanced facade with pedimented portico and fanlight over the front door. A few Greek Revival-style houses exist in Milledgeville, such as the Thirteen Columns House on Wilkinson Street, built about 1825. Most houses of historic interest are private residences; however, two are included in the Trolley Tour, which leaves twice weekly from the Chamber of Commerce.

The *Marlor House,* on Wayne Street, was built in transitional Federal style in 1830 by John Marlor, a master builder-architect. Many of the fine homes of the area of the 1820 to 1830 period are attributed to him, including the second house open for public inspection, the *Stetson–Sanford House,* ca. 1825. This building is distinguished by a Palladian-inspired double portico, with Doric columns on the first floor and Ionic columns on the second. The *Masonic Hall,* built in the early 1830s, was also designed by John Marlor with Italianate detail. An especially noteworthy feature is the eighty-seven-foot unsupported circular stairway extending three flights.

The *Old State Capitol,* now Georgia Military College, located in Capitol Square, is one of the oldest public buildings of Gothic Revival style in the United States. The central part of the building was erected in 1807; wings were added in 1828 and 1837. Here Georgia enacted its secession ordinance of January 19, 1861. When Milledgeville was occupied by Sherman's left wing, in November 1864, the capitol was extensively vandalized. The impressive three-arched gates on Green and Franklin streets were built in late 1865.

Of all the fine buildings in Milledgeville, the most noteworthy architecturally is the *Old Governor's Mansion,* built in 1838 and occupied by Georgia's governors until 1879. This Greek Revival mansion, a National Historic Landmark, is notable for its impressive Ionic portico, central rotunda with coffered domed ceiling and skylight, and architrave frieze wrapping around the house in a symmetrical fashion. The design is attributed to Irish-born Savannah architect C. B. Cluskey, one of the most important architects in the Greek Revival movement. This magnificent mansion cost $50,000, a large sum for that time and over three times the amount originally appropriated. Construction is of brick with stucco overlay, popular at the time. The fence and accessory buildings were built later by convicts from the state penitentiary.

The mansion was restored and renovated in 1967, and now houses the offices of the president of Georgia College, a unit of the Georgia university system. The mansion had been donated in 1889 to attract to Milledgeville the Georgia Industrial and Normal College. This institution was the predecessor of Georgia Women's College, which became Georgia College in 1976 and is now coeducational. The college now occupies the original state-penitentiary square, burned in 1864 while Union troops were in the city. Of particular interest on the campus are the Ina D. Russell Library's Flannery O'Connor Room, which contains memorabilia from Andalusia, the country estate where O'Connor lived, and the Museum and Archives of Georgia Education.

HISTORY OF MILLEDGEVILLE AREA

The development of Milledgeville as the state capital in 1803 is part of the story of Georgia's great obsession with the expulsion of the Indians and westward expansion into the rich, brown loam soils of the Piedmont.

Throughout the Revolutionary War, thousands of Virginians and Carolinians poured into the newly opened lands in Wilkes County. Thus, until the end of the war, there existed two Georgias: an up-country, land-hungry group of settlers who espoused the Revolutionary cause, and a low country held by the king's troops, with the colonial capital at Savannah.

The 1783 Creek land cession effectively moved the westward boundary of settlement from the Ogeechee to the eastern bank of the Oconee, and the boundary was confirmed in the Treaty of New York in 1790. But Georgians had no intention of thus restricting settlement. The trans-Oconee area became a source of contention, with federal forts along the Oconee at such critical crossings as Rock Landing, 6 miles south of the site of future Milledgeville, to keep whites from intruding on Creek territory. The result was a tenuous peace, with white incursions into Indian land and Creek depredations on white settlements.

Indians became increasingly dependent on trade goods from whites, and the old story of land cession to pay off trade debts was repeated. By 1802, the Creeks ceded the land west to the Ocmulgee. An important part of the treaty amendments of 1806 provided for the construction of a Federal Road from the Ocmulgee River to Mobile. This was opened almost immediately, and became an important settlement route to the Gulf area, assuring that Americans settled land in West Florida then disputed with Spain.

The 1802 cession marked the introduction of the land-lottery system, outlined in an act of the Georgia legislature on May 11, 1803, which created three new counties. Surveyors laid out tracts of forty-five chains square, parallel to the dividing line between Baldwin and Wilkinson counties. Thus the original land lines ran from northwest to southeast and northeast to southwest. The land thus surveyed was distributed by a lottery of 1805, held at Louisville; six other land lotteries were all held in Milledgeville.

The act of 1803 also provided for the location and survey of a town called Milledgeville. This tract would contain sixteen lots of 202.5 acres (forty-five chains square), for a total of 3,240 acres. Only a small part was to be residential, with one-acre lots. The town plat of 1804 contained only 500 acres, but showed four public squares of 20 acres each. The town streets were laid out at a 45° variance with the original survey lines. In 1804 the legislature made the embryonic town of Milledgeville, near the geographic center of the state, the capital. Houses in Milledgeville were set close to the street, thus providing more space in the rear for garden, orchard, poultry house, and other outbuildings.

In 1807 work on the new statehouse had advanced sufficiently for the legislature to meet in Milledgeville. It last met here in 1866, when Military Reconstruction authorities transferred the balance of state power to Atlanta. Despite being the state capital, town growth did not equal that in other parts of the state. In 1860 Milledgeville was tenth of the sixteen cities in Georgia with a population of 1,000 or more.

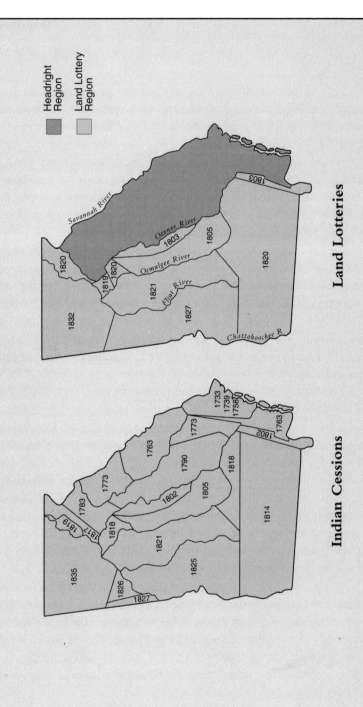

One deterrent to growth was Milledgeville's relative isolation and lack of transportation facilities. Before the Civil War, most of Georgia's roads were hardly more than wagon trails. The state improved navigation on the Oconee, but even after the introduction of steamboats, freights remained high. In 1836, at a time when rates per hundred pounds between New York and Darien, Georgia, were $2.37, the cost from Darien to Milledgeville was over a dollar—and it took a week to make the trip.

As the years went by, there was some growth in state services in Milledgeville. The state penitentiary became operative in 1817 and by the 1830s inmates were engaged in twenty different crafts. Selling on the open market in the penitentiary store was much resented by craftsmen in the town. The penitentiary was burned in 1864, and became unimportant in the life of the town with the institution by military governor Thomas R. Ruger in 1868 of the convict lease system, for which the state treasury received a few pennies a day per convict.

One of the largest employers of the area today is the Georgia State Hospital, the central mental institution of the state. It was authorized by the legislature in 1837 as the Lunatic Asylum of the State of Georgia and began operation in 1842. The main central building, known as the Powell Building, was completed in 1856 in Greek Revival style.

Milledgeville was the Georgia capital throughout the Civil War. Its first actual contact with fighting, however, did not come until the Stoneman Raid of July 1864, when a group of federal cavalry burned the depot and destroyed rolling stock at Gordon before moving on to Milledgeville. When the left wing of Sherman's army entered Milledgeville on November 20, the city earthworks had been abandoned and Governor Joseph Brown and the legislature had fled. There was extensive vandalism of property—burning of fences and outhouses, stabling of horses in churches—but no great burning of private property. The Central Depot, the bridge over the Oconee, and the state arsenal were burned and the fireproof magazine blown up. The Oconee Mill, the textile factory, and the foundry escaped destruction, probably because the owners were Northerners or of foreign birth. Most of the large homes were occupied by Union officers, and thus received some protection. General Sherman himself occupied the governor's mansion.

On June 29, 1865, James Johnson, a well-known Unionist before the war, was appointed provisional governor of Georgia by President Andrew Johnson. A constitutional convention was held in October at the statehouse and, except for the exclusion of a few high-ranking Confederate officers who had received a pardon but could not vote, voters at this convention were the same as those who voted for secession delegates who took Georgia out of the Union. The Negroes, now called freedmen, took no part in these 1865 political activities. Under this constitution of 1865, Charles J. Jenkins was elected governor and was inaugurated on December 14.

The capital was in disarray. The change in the status of the Negro, and the complete absence of money—Confederate currency was worthless—made it extremely difficult to accomplish the simplest jobs. The small federal garrison of twelve men tried to be helpful and assist in finding work for former slaves, but with little success. The most active agency in the town during this early Reconstruction period was the Bureau of Refugees, Freedmen, and Abandoned Lands. Popularly known as the Freedmen's Bureau, it acted as an educational and social relief agency, opening schools and supplying food to both blacks and whites as well as settling local wages and affixing conditions for working under contract. None of the local officials were carpetbaggers or scalawags. Thomas W. White was a lawyer, whose father had been one of the county's largest landowners in 1860. Matthew R. Bell, a native of Forsyth County, had served as an enrolling officer with the rank of captain in the CSA. The Bureau helped in the transition from a slave society to a free society in a very difficult period, although it had very limited resources.

Milledgeville's improved conditions were to be short-lived. Threatened with possible loss of a majority in Congress, Radical Republicans scrapped the reconstruction policies of President Johnson, and initiated a policy of Military Reconstruction under the supervision of a Joint Congressional Committee. When Georgia rejected the Fourteenth Amendment in November 1866, it was placed under Military Reconstruction. A new registration of voters was made, and elections held for a new constitutional convention. The 1867 convention was not held in Milledgeville, but in Atlanta on orders of General John Pope. The alleged grounds for removal to Atlanta was the refusal of the innkeepers in Milledgeville to house the thirty-seven Negro delegates elected to the convention. This de facto move by military fiat was confirmed in February 1868 when a clause was inserted into the new constitution making Atlanta the state capital.

With the beginning of Radical Reconstruction, the Freedmen's Bureau became an increasingly political organ helping Republicans control the state with the Negro vote, and this sullied the reputation it first had. The agency lasted only until 1870; and the greatest expectation of the free Negro, to receive an allotment of land in fee simple—"forty acres and a mule"—was not achieved.

With the compromise of 1876, in which southern Democratic electors voted for Republican presidential candidate Rutherford B. Hayes instead of Democratic candidate Samuel Tilden, the North lost interest in the condition of the Negro in the South. In the period that followed the Negro became a day laborer or sharecropper in a continuation of a decentralized plantation system. With a decline of the price of cotton to as low as four cents a pound in the 1890s, all of the South, black and white, suffered the effects of deep poverty.

At the end of the Reconstruction period, secessionist fire-eater Robert Toombs led the move for drafting a new state constitution, although there was no great enthusiasm in urban Atlanta for replacing the Constitution of 1868. Milledgeville had high hopes that the new document would provide for restoration of the state capital to Milledgeville. However, when the question of the capital of the state was put on in a referendum, separate from the Constitution of 1877, Atlanta remained the capital by a substantial majority: 43 to 934. Georgia was the only Southern state to have its capital changed as a result of Reconstruction.

WILKES FLAGG

One of the most remarkable African Americans to live in Georgia during the antebellum and Reconstruction periods was Wilkes Flagg of Milledgeville. His career brought into sharp relief the many contradictions of black-white relations which existed in the Old South. He demonstrated how perseverance could bring prosperity to black men within a slave society and how his wealth and leadership could be used to further the progress of black citizens.

Wilkes Flagg, father unknown, was born in Virginia about 1802, the son of Sabina, a slave. In 1831 he was bought by Tomlinson Fort of Milledgeville, a prominent physician, politician, landholder, and railroad promoter. Fort had a paternalistic attitude toward slavery, though he never manumitted a slave. He permitted his children to teach Flagg to read and write and to keep accounts, and Flagg was allowed to use the income from his work as a blacksmith after regular working hours to purchase the freedom of his wife and son. However, because of the equivocal success of free men of color in Georgia, he technically remained the slave of Fort, who lent his name to loans made for the benefit of Flagg. When the Civil War began, Flagg had accumulated an estate worth over $25,000, a respectable sum for those hard-dollar days, and his wife Lavinia lived unusually well for an African American in those times of slavery.

Early in his life, Flagg had acquired the skills of a waiter; and it was said that he was much in demand for state dinners given by Milledgeville governors from Wilson Lumpkin to Joseph Brown. According to tradition, Flagg was a handsome man, "copper colored, six feet tall, with the manners of a Chesterfield," with good taste in dress and precise speech. Flagg was personally an abolitionist, but he accommodated

himself to the prevailing white milieu by never expressing these views publicly or from his pulpit as a Baptist preacher.
 With the close of the Civil War, Flagg's fortune declined along with that of his white neighbors, but he continued to prosper. Blacksmith skills continued to be much in demand. Of a conservative social bent, he believed it to be a great injustice to free the blacks without educating them to the responsibilities of their new status and helping them economically. He bought land on which to establish a freedman's colony and he built Flagg's Chapel to educate former slaves in social and economic as well as religious matters. When the freedmen were given the vote, Republicans encouraged him to enter politics. He declined and, as a result, some of his chapel members broke away to form rival Hamp Brown Church.
 Flagg continued to work hard for himself and his people; but old age found him financially insolvent. On his death in 1878 he had only a house to leave to his widow. After her death, she and her husband were interred under the floor of the chapel that Flagg had constructed.

Milledgeville to Macon, Ga. 22, U.S. 129, 35 miles

Leaving Milledgeville, follow Ga 22 southwest to Gray, the country seat of Jones County. Carved out of Baldwin County in 1807, the county was named for James Jones, a prominent Savannah attorney and congressman; the town took its name from James J. Gray, one of the major financiers of the Confederacy.
 Gray was just a hamlet on the Central of Georgia when in 1905, Jones County citizens voted to move the courthouse to the railroad town. The Romanesque-style painted brick courthouse is on the left, just after crossing the railroad tracks, when going south on US 129.

✓ On to the south is *Clinton Roadside Park*. Here huge granite outcroppings in a wooded area mark the Piedmont fall line. Historic markers give details of the significance of Clinton in early Georgia history.

Listed on the National Register of Historic Places, Clinton was the county seat of Jones County for almost a hundred years (1808–1905). Until 1850 it was a major manufacturing center for cotton gins. That year Samuel Griswold transferred his factory to Griswold, about nine miles east of Macon on the newly completed Central of Georgia Railway. Of three Clinton buildings included in the Historic American Buildings Survey, only the *Mitchell–Barron House* (called Barron-Blair House in the 1975 Clinton, Georgia, guide published by the Old Clinton Historical Society) is still standing. This house is a two-story L-shaped house with two porches; the first floor had Doric columns; the second floor, Ionic. Twelve antebellum buildings remain in Clinton's central area; the restored McCarthy–Pope House, dating from 1810, is the oldest. The site of the Clinton Female Seminary is now the Old Clinton Historical Society's museum and visitor center.

Clinton suffered heavily in the July 1864 Stoneman Raid; and it was here the main body of raiders, including Union Major General George Stoneman, was captured. Four months later, Clinton found itself in the path of Sherman's right flank; the town was sacked and a third of the houses burnt. Griswold's plants, both the buggy factory in Clinton and the gun factory at Griswoldville—the cotton-gin factory had been converted to the manufacture of revolvers—were in ruins. Griswold died in 1867 and was buried in the Clinton Cemetery behind the Methodist church, which dates from 1822. An impressive marker and fence marks the grave of this early Georgia industrialist of the Plantation South.

Old Clinton, a booming frontier town before Macon was laid out, is a reminder of what happens to cities when transportation technology changes. Clinton lost out twice when railroads built to serve the cotton-growing areas of Georgia bypassed it. In the 1840s the Central of Georgia reached the Ocmulgee 12 miles to the south at Macon; and in the 1900s, a new line was pushed just a mile and a half north of the Clinton courthouse.

Continue on U.S. 129 to Macon.

Cotton and the South

Cotton, indigenous to Asia and the Americas, is one of the most widely diffused fiber plants in the world. Long cultivated, spun, and woven in India, cotton was probably introduced to Europe by Alexander's invading forces. When the Spanish arrived in the New World they found cotton extensively used by Indians in Mexico, South America, and the Caribbean.

COTTON GROWING IN BRITISH COLONIAL AMERICA

From the earliest days of colonization, there are scattered accounts of growing cotton for domestic purposes. Both green-seed (lint adhering) and black-seed (lint free) varieties were grown. A preference for the black-seed cotton was reinforced by the introduction of roller gins, modelled after the treadle-powered *chur ka* used in India. With an improved model of the device even a boy could turn out seventy to eighty pounds of lint cotton a day, thus eliminating the laborious task of separating the fleece from the seeds by hand, a nightly task for slaves.

Records show shipments of cotton to England, as well as of intercolony shipments, before the Revolution. However, during the colonial period, cotton was grown mainly as a patch crop for domestic use. Cotton diseases and insects are noted early; the "cotton weevil" was reported in 1728 in North Carolina.

Cotton as a money crop was largely a nineteenth-century development, by which time the combined technologies of spinning and weaving helped fuel an insatiable demand in industrializing England for cotton. The plantation system, first in South Carolina and then spreading west to Texas, embraced upland cotton as its cash crop. Cotton soon came to dominate the political as well as economic fortunes of the South and helped to perpetuate the demand for slaves, this at a time when the rest of the United States was moving toward industrialization and urbanization.

64   THE PLANTATION SOUTH

Both this and the photo on the following page could have been taken any time between 1900 and 1930. Here the field hand is emptying picked cotton into split-oak baskets.

COTTON VARIETIES, CLIMATE, AND TECHNOLOGY

Cotton is identified by the area where it is grown, the color of the seeds, and its fiber length (staple). The fleece of green-seed cotton adheres tenaciously to the seed, while the fiber of black- or smooth-seed cotton can easily be separated from the seed by means of a roller gin. Both varieties were cultivated in America in the colonial period. After the Revolution cotton exports began with Sea Island cotton, a variety that could only be grown in a limited geographic area. Always a specialty crop, it required much care in preparation for ginning and marketing, and only accounted for a very small part of the American cotton crop.

The term "upland cotton" was applied to many kinds of short-staple cotton whose cultivation was well suited to the heavy meta-

Baled cotton is ready to be loaded onto ships, probably in Savannah, Georgia.

morphic-derived soils in South Carolina and Georgia above the sand-hill divide that separated the Coastal Plain from the Piedmont. Later, with the introduction of commercial fertilizers, short-staple cotton moved into the Coastal Plain; but the "Old Cotton Belt" in both South Carolina and Georgia was originally in the Piedmont. There cotton displaced tobacco as a backcountry commercial crop and hence the term "upland" got attached to short-staple, green-seed cotton which could be grown in these areas. Short-staple cotton was adapted to a wide variety of growing conditions and did not require painstaking care in picking, cleaning, and molting.

Two technological innovations contributed to increased cultivation of short-staple cotton: the saw gin and a new hybrid variety. Eli Whitney invented the bent-teeth cotton gin in 1793 at Mulberry Grove Plantation, near Savannah, and also took out a patent for a circular saw gin. However, this improvement is also variously attributed to Hogden Holmes and Edward Lyon.

Whitney hoped to maintain a monopoly through toll ginning, but the idea of the gin was picked up quickly by country mechanics and soon had spread as far west as Mississippi. The effect was immediate. In 1794 the United States produced only 8 million pounds of cotton, mostly of the Sea Island type. In 1804, production reached 64 million pounds; in 1811, it was 80 million, with South Carolina producing 40 and Georgia 20 million pounds.

Of equal importance with the new gin was the introduction of Mexican cotton and its hybridization with the upland varieties, which had degenerated from lack of attention to breeding and were difficult to pick and dirty to gin. The boll of the Mexican cotton opened wide, which permitted faster and cleaner picking and thus cleaner ginning.

SEA ISLAND COTTON

Cotton production along the Georgia and South Carolina coast began as an alternative to indigo production, unprofitable when the British government ceased to foster its production after the Revolution. There is no agreement on how improved strains of cotton were first introduced—but it is likely that they came from the Bahamas. The production of Sea Island cotton was eagerly taken up by a number of planters because it brought premium prices on the English market.

However, although Sea Island brought twice as much as upland cotton, the price advantage often disappeared with the greater costs in preparing it for market.

Although generally identified with the off-shore islands, Sea Island cotton expanded into coastal areas suitable to the production of black-seed cotton. The cotton was well suited to the sandy, light yellow soils of the coast, but needed constant manuring with marsh mud and compost to maintain soil fertility for good yields. For best results, forty cartloads were applied to each acre in the winter. This hard, dirty work was hated by hands, and sometimes neglected by slack masters, with a consequent decline in yield.

Sea Island cotton production was labor intensive and required constant attention to turn out the finest quality staple. With care, Sea Island production brought a 10 to 12 percent return on investment; but such a handsome return was the exception rather than the rule.

Unlike short-staple cotton, which transformed the plantation and politics of the South, Sea Island cotton made no great changes in the low country economy. It was but a minor staple in the antebellum period, accounting for less than 2 percent of U.S. cotton production in 1858. This cotton is no longer produced commercially in the United States.

SHORT-STAPLE COTTON

Improved seeds and production technology arrived at a time when the Piedmont areas were being opened up to white settlement. As the Indians were moved westward ahead of the encroaching white tide, new land was always available for cotton growing. In eighteenth-century Georgia and South Carolina, upcountry agriculture had always been sharply differentiated from that in the low country, an area of large plantations and many slaves in proportion to the white population. In contrast, slaves were few in the upcountry where yeoman subsistence farmers followed a mixed agriculture based on animals and grains and local, diversified handicrafts. The most important cash crop around Augusta was tobacco; but its sale or barter was primarily to achieve necessities and a few luxuries.

The introduction of short-staple cotton as a cash crop changed the agricultural emphasis and had a profound impact on the culture of the South, for it spread slavery. Despite a steady increase in the population base, the proportion of slave population increased steadily in up-country South Carolina counties. In many the slave population had become a majority by 1840. This demographic change had profound political consequences. During the eighteenth century in the South, there had been a political cleavage in interests between low country and backcountry. The slave economy of short-staple cotton amalgamated, in economic and political outlook, the Piedmont with the older coastal areas.

Investment capital was diverted from nascent agribusinesses, such as iron works and small water-powered mills, to cotton growing and slaves. Grain raising and livestock raising were neglected, and by 1843 Camden and Columbia, former exporters of grain and flour, had become importers of Northern grain and hay. The self-sufficient pioneer homestead and plantation both gave way to "store bought" goods which the industrializing states were more than happy to provide. The cotton-growing South shipped cheap cotton to England and the North and bought back shoddy goods. The trend became even more pronounced after the construction of railroads. First built to facilitate the shipment of cotton out of the South, the railroads were eager to fill their empty inland-moving cars. Relatively cheaper rates on manufactured and imported goods contributed to the decline of plantation industries.

The most limiting factor in the production of cotton continued to be the necessity of picking by hand. At the beginning of the nineteenth century, a hand could pick only about 50 pounds a day from small, tight bolls. By 1860, wider, fuller bolls increased that to 150 to 200 pounds a day—an amount not exceeded except by exceptional pickers for the next hundred years. Only after World War II were successful mechanical cotton harvesters introduced. By that time, cotton was already in decline in South Carolina and Georgia.

Regional differences in soil fertility gave cotton producers in Mississippi and Texas a great advantage over those in South Carolina and Georgia. The princely profits of the early cotton days quickly declined in the aftermath of the War of 1812. Cotton prices rebounded in the 1850s, but the increase did not reflect substantial returns on investment.

The increasingly adverse terms of trade for Old Cotton Belt planters is one reason that the election of Abraham Lincoln and the triumph of the Republican party was perceived as a threat sufficient to merit secession.

Short-staple cotton, on which the plantation economy of the South relied, was also responsible for a political-economic climate that ended in secession as a way of solving problems in an economy increasingly tilted in favor of the industrial states of the North. To the Southern farmer of the 1860s, it seemed that everybody who in any way handled cotton—from the factor to the buyer, to the insurance broker and railroad carrier—made money off cotton except the men and women who grew it on the land. This condition continued throughout the twentieth century, until finally the Old Cotton Belt, which had survived the boll weevil and the Depression, gave up growing cotton.

Macon, Georgia

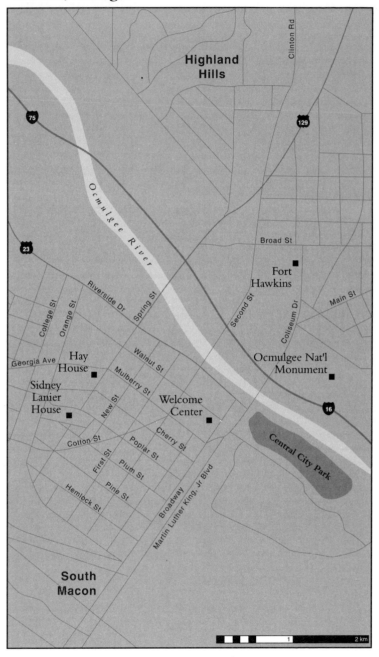

⃤ *Day Four*

MACON AND VICINITY

Located on the Ocmulgee, Macon is the county seat of Bibb. It is one of three cities located along the fall line in Georgia that have had continuing modest growth; the other two are Augusta on the Savannah and Columbus on the Chattahoochee. Formerly an important middle-Georgia terminal in the heart of the cotton belt, Macon suffered heavily after World War II with the decline in agriculture, cotton manufacturing, and rail traffic. Shops and businesses fled the downtown area and in the 1960s Macon took on the look of a dying rural town, magnified in scale. Historic preservation, downtown rejuvenation, and the vigorous pursuit of new industry has produced a dramatic turnaround. Symbol of the new Macon is the headquarters of the Cherry Blossom Festival and Keep Macon-Bibb Beautiful Commission at the corner of Cherry and New streets. This impressive Georgian-style building that blends with the surroundings of downtown Macon was made possible by a grant from Yoshida Kogyo Kabushkigaisha. Yoshida opened a plant to make zippers in Macon in 1974; the plant now accounts for over half the domestically manufactured zippers sold in the United States.

Points of interest on a Macon tour include attractions outside the city: Indian mounds, a rebuilt nineteenth-century blockhouse, and a complex of farm buildings, some of which date from 1830; within Macon, there are stops at the Hay House, Sydney Lanier Cottage, and Harriet Tubman Historical and Cultural Museum. Additional information about historic sites, tour guides, and maps are available from the Macon-Bibb County Convention and Visitors Bureau, located in the Terminal Station on Cherry Street. It is located in what formerly was the passenger station for the Central of Georgia Railroad.

Fort Hawkins

The reconstructed blockhouse at Fort Hawkins is a mile north of the Terminal Station on Emery Highway (US 23), across the Ocmulgee. The fort was named after Indian Agent Benjamin Hawkins, who was responsible for all Indian affairs south of the Ohio. It was built in 1806 after the Creek land cessions extended the boundary of white settlement beyond the Oconee. The fort consisted of a 100-acre reserve with blockhouse and stockade, and it served as agency headquarters and trading post and warehouses for the Indian trade.

During the Creek Wars (1812–1814), which were a by-product of the U.S.-British War of 1812, Hawkins was an important point for the rendezvous and disposition of militia. The last eventful year for the fort was 1817: General Andrew Jackson, with a large contingent of Tennessee militia, arrived enroute to Florida to fight the Seminole; and, on another occasion, an Indian Assembly of 1,400 who gathered to receive their annuity for cession of land.

With the Treaty of 1821, the Creeks formally ceded the land between the Ocmulgee and Flint rivers. The Creek Agency was moved to the Flint River, and the brief history of Fort Hawkins came to an end. Macon was laid out on the west bank of the Ocmulgee.

Ocmulgee National Monument

Just to the south is Ocmulgee Old Fields, on the Macon Plateau, one of the most important Indian archaeological sites in the United States. Its history of human occupation spans ten thousand years. Excavated under auspices of the Works Progress Administration during the 1930s, Ocmulgee is administered by the National Park Service. Visitors are cautioned to park only in the designated spaces, and to walk only on the footpaths. The visitors center has a major archaeological museum. The *Earthlodge*, a restored ceremonial building, may be reached by a walking trail from the visitors center. The Great Temple Mound, early *English Trading Post* site, and *Funeral Mound* are accessible by car and walking trail. Admission is free.

Indians of Georgia and South Carolina

The Indians of Georgia and South Carolina belonged to the Southeastern Culture Group characterized in the Mississippian period by a religious and world view which had striking analogies with the more-developed Indian civilizations of Mesoamerica. Disease and social disorganization brought on by sixteenth-century intrusions of the Spanish brought an end to the large chiefdoms of this period. Tribes of the historic period, such as Cherokees and Creeks, reflected the amalgamation of Indian groups forced to coalesce for survival after the collapse of Mississippian culture.

PALEO-INDIAN PERIOD (TO 7,000 B.P.)

Hunters of big game entered the Americas from Asia across the Bering Strait, then a land bridge when much of the world's water was locked up in the glaciers of the Pleistocene. The time of entry had been estimated from 20,000 to 50,000 years ago. By 7,000 years ago, they had become widely distributed throughout the North and South American continents. They hunted Ice Age animals, now extinct. Except for lance points similar to those found in Siberia little is known about Paleo-Indian culture. No sites have been identified in Georgia, and throughout the Southeast there is little evidence of these big-game hunters. The first known trace of humans on the Macon Plateau consists of a fluted stone spear point of a type dated about 11,000 years ago in other parts of the United States.

ARCHAIC PERIOD

As the climate became warmer and drier, the nature of the flora and fauna changed, and the hunter of big game merged into the generalized hunters and gatherers of the Archaic period. They hunted a great variety of small game and birds, gathered wild plants, fished the streams and lakes, and gathered clams and mussels. The shell midden on Stallings Island, in the Savannah River near Augusta, is from this period. Social organization consisted of small nomadic bands; but site exploitation became more systematic.

More tools were introduced and artifacts now included stone axes for working wood, smoking pipes, earthen pots, woven baskets, and ornaments of copper and shell, indicating the occurrence of trade between Southeastern and Northwestern groups. Even more important was the introduction of burials with artifacts, which indicate the elaboration of a spiritual as well as material culture.

THE WOODLAND PERIOD

Associated with the beginning of agriculture in the Southeast, the Woodland period began on the Macon Plateau about 3,000 years ago. Agriculture diffused rapidly throughout the eastern wooded area of what is now the United States. Gourds and squash, and later corn and beans, were introduced. Gardening, rather than farming, best describes this early agriculture, and hunting, fishing, and gathering continued to be important food sources.

Corn was probably introduced from Mexico where it was grown for food 5,000 years ago. Perhaps the earthen ceremonial mounds of this time also had roots in Mesoamerica. Diffusers of these ideas to the Southeast Indians may have been *pochtecas,* traveling traders, whose presence in North America is supported by the curious shell and copper masks of their god Yacatecuhtli found in many Mississippian sites.

The Indians now lived in larger, more permanent groups for much of the year, although they split into smaller groups for fishing and gathering. The use of pottery increased in quantity and quality, with decorated designs stamped in the damp clay. Burial beneath rounded earthen mounds spread widely, and valuable and elaborate ceremonial goods were interred with the dead. While burial mounds are found in the general area, none were found within the present confines of the Ocmulgee National Monument.

THE MISSISSIPPIAN PERIOD

The Mississippian period came to the Macon Plateau about 1,100 years ago and ended at the time of European contact. The Mississippians practiced extensive riverine agriculture, based on corn, beans, and squash; and founded larger, more stable towns with spheres of influence over satellite settlements. Descriptions of Hernando DeSoto's traverse indicate the Indians of the Southeast were constructing new mounds and increasing the size of old ones when he made his destructive trip of discovery in the 1540s.

From the early Mississippian period there is evidence of a palisaded town with a double row of surrounding trenches; rectangular houses of poles set upright in the earth, plastered over, and roofed with thatch; a large underground Earthlodge or council house; and a number of ceremonial mounds. Sites in Georgia and South Carolina indicate a much larger Indian population than was present at the time of European contact.

For unknown reasons the Macon Plateau was abandoned after about 200 years' occupation. The settlement discovered 3 miles south of the Macon Plateau in a swampy area was founded later and referred to as Lamar (late Mississippian). This village contained two temple mounds and was surrounded by a stockade. These early Georgia Indians were probably the ancestors of the fifteenth-century Creek, although Creek tradition has their ancestors arriving from the west.

HISTORIC PERIOD

The first Europeans arrived in the Southeast with the landing of Ponce de Leon on the east coast of Florida in 1513. Much of our early information and misinformation about Indians of the Southeast at that time is derived from Francisco de Chicora, an Indian captured in 1521 by a slaving party in South Carolina and taken to Spain. Attempts at colonization by Europeans were ill fated until Pedro de Menendez founded St. Augustine in 1565. The English planted their first South Carolina colony, Charles Towne, on the Ashley River with the aid of the friendly Kiawah—a tribe of the coastal Cusabo. Charles Towne immediately launched an Indian trade in both deer skins and slaves, using the Westo on the banks of the Savannah as slavers, and in turn destroying them in their pursuit of a more lucrative trade to the west. By the end of the century, Charles Towne traders had pushed to the Mississippi, but their importance has been largely neglected by historians. Charleston traders made the Indians dependent on European trade goods, for which they eventually bartered away their land at discount prices.

Because Ocmulgee Old Fields lay astride the historic Lower Creek Path, it very early became an important trading point with the Creek Indians. As early as 1690, the Charles Towne merchants had established a trading post at Ocmulgee. To assure control of the western trade, the Carolinians decided to destroy the flourishing Apalachee towns that lay under Spanish protection midway between St. Augustine and Mobile. Continued Carolina attacks over the years eventually decimated not only the Apalachee but the flourishing Timucuan settlements in North Florida. Within a few years the missionary efforts of the Spaniards, which had extended over a hundred years, were obliterated.

As long as there was an effective French presence, the Indians were able to take advantage of competing French, Spanish, and British interests. With the Treaty of Paris of 1763, and the establishment of English hegemony, the Indians of the Southeast lost this. Even before the Revolution, large numbers of settlers had been moving along the Appalachian front southwestward into South Carolina and Georgia. These settlers had a single economic and political objective: to get rid of the Indians so they could occupy the land. In South Carolina, many Indian groups had already become extinct. Early South Carolinians had depended on Indians for food for survival and trade for prosperity. The

Indians' reward was disaster: they were killed in wars, captured and sold as slaves, dead from disease, absorbed on the fringes of white settlement, or forced west to merge with larger Indian groups.

But in Georgia the case was different. Georgia claimed land west to the Mississippi, a princely domain that included Cherokee, Creek, Choctaw, and Chickasaw land. In 1802, an agreement was negotiated by the state with the U.S. government in which lands west of the Chattahoochee were ceded for $1.25 million and a promise to extinguish Indian claims to land still held by them in Georgia. Until the forcible military expulsion of the Cherokees in 1835, politics in Georgia were dominated by the question of Indian lands: at the federal level, in pressuring the United States to get all Indians out of the state.

In the process of Indian removal, little thought was given to the fact that the Creek and Cherokee Indians had been drastically changed by two hundred years of contact with the whites. By the time of removal, most were small farmers with a pattern of living and subsistence akin to that of the whites pushing on their frontiers. They lived in the same kind of log cabins, used the same tools, wore similar dress (except Indian men still preferred the long tunic shirt), ate similar food, and practiced a basic scratch farming based on the deadening of trees and the clearing of new ground. This process of acculturation, most advanced among the Cherokee with their surge toward literacy through Sequoyah's alphabet, was interrupted by forced removal.

Removal delivered the formerly self-sufficient Southeastern Indians into a condition of crowding and poverty in the Indian Territory west of the Mississippi, where they were restricted to reservations. Deprived of the lands needed to maintain their traditional existence and forced to live in enclaves as wards of the government, the Indians were converted from a once proud and independent assemblage of people into a poor and dependent group which could live neither as Indians nor as Americans. But U.S.-Indian relations have not been unique; rather they must be viewed as a pattern in which peoples of European origin have subjugated and exploited peoples of other races, whether in South America, Africa, Asia, Australia, or Oceania.

Macon Historic District

The Hay House is a twenty-four-room Tuscan-style villa, built between 1855 and 1860 by Macon industrialist–entrepreneur William Butler Johnston. It is the largest and most ornate antebellum residence in Georgia in a style rarely seen in the South at that time. William B. Johnston (1809–1887) moved to Macon in 1831, only eight years after the city was laid out. An energetic businessman and astute promoter, he soon branched out from a lucrative jewelry business into real estate, cotton, ice making, gas- and water-works, and life insurance. A major source of wealth was investment in railway stocks. In 1851 he married Anne Clark Tracy, daughter of Macon's first mayor.

On an extended honeymoon in Europe, the Johnstons came under the spell of Second Empire architecture. They brought back from Italy artisans and many materials to aid in the construction of this elegant villa. The formal Italian gardens which once covered the entire block are now gone, but the house retains many features far in advance of Georgia buildings of its day—central heat, indoor plumbing, a spring-fed water system, a cold-well summer ventilation system, and a coal lift. Johnson bought many paintings and pieces of sculpture for display in the largest room in the house, the "picture gallery."

Parks Lee Hay, founder of the Bankers Health and Life Insurance Company, acquired the Johnston property in 1926, and filled the villa with antiques and other valuable furnishings collected on trips abroad. In 1977 the Hay family gave the house and its furnishings to the Georgia Trust for Historic Preservation, which since then has operated the Hay House as a museum. Children in particular enjoy being shown the "secret room," accessed by the niche on the main stair landing which actually is a concealed door. Legend has it that part of the Confederate gold was stored in this room during the war.

The ornate Hay House and its furnishings provide an interesting contrast to the simpler furnishings and more restrained interiors of Greek Revival architecture popular in Macon and which are characteristic of such antebellum homes as the *Woodruff House (Overlook)* and *Old Cannonball House,* both open to the public, and the Holt and Nisbet houses, which are private residences.

Lanier Cottage is the birthplace of Georgia's best-known poet Sidney Lanier (1842–1881) and presently the headquarters of the Middle Georgia Historical Society. The simple story-and-a-half gable house with one central dormer derives its importance from the poet. Lanier's poems, written between 1869 and 1871, utilized the Cracker dialect, e.g., "Thar's More in the Man Than Thar Is in the Land." Later he switched to nature-inspired poetry, among which "The Marshes of Glynn" is one of the best known. At a time when much of the South followed the get-rich industrial creed, Lanier adhered to a belief in an agrarian ideal of a diversified agriculture and sturdy yeomanry, black and white, living independently and tilling the soil. His essay "The New South" restates the theme of many of his poems—the salvation of the South lay in its wise use of the land, a struggle in which contemporary Georgia environmentalists are engaged today.

The Harriet Tubman Historical and Cultural Museum is named after the most famous guide on the Underground Railroad which helped runaway slaves get to safety in Canada. Harriet Tubman had escaped to freedom from a Maryland plantation in 1849, and in 1850 began a ten-year career leading some 300 slaves to freedom. During the Civil War she worked for federal forces on the South Carolina coast as laundress, nurse, and spy; but when she died in Auburn, New York, in 1913 she had never received a pension for her services.

The Macon Harriet Tubman Museum was founded in 1982 by a diverse group of local citizens and was opened in 1985. With permanent exhibits depicting the history of blacks in Africa and America and visiting exhibits concerning contemporary cultural contributions, this museum adds significantly to Macon's cultural resources.

Jarrell Plantation Historic Site

This authentic middle Georgia farm located 25 miles northeast of Macon began as a cotton plantation in the 1840s and survived the Civil War, Sherman's March to the Sea, Reconstruction, and the Depression. It was still a working farm into the 1960s, when Willie Lee Jarrell, grandson of John Jarrell, plantation founder, retired. The Jarrell family donated the homeplace and the surrounding twelve acres to Georgia to become a State Historic Site, and it has been open to the public since 1976.

The plantation's twenty buildings stand where they were built—between 1847 and 1945. Fifteen of these buildings are on the National Register of Historic Places as fine examples of Middle Georgia farm structures; and they contain one of the largest collections of original family artifacts in the country. Buildings include

Plain-style plantation house on the Jarrell Plantation Historic Site. John Jarrell built the house in 1847. It contains the original furnishings.

the plain-style plantation house, with original furnishings, built in 1847 by John Farrell and a smaller house built in 1895 by Jarrell's son Benjamin. The third "big" house visible from the walking trail was built in 1920 and is not open for visits. The mill complex includes cotton gin (1895), sawmill (1895), planer (1917), shingle and cane mill (1916), and wood-burning steam boiler (1909). The blacksmith shop has an adjoining carpentry shop, which houses tools of the tanner, wheelwright, and cooper, plus tools for shingle-splitting and log hewing.

A self-guided tour begins at the visitors center, where a seven-minute film provides an overview. A numbered pathway leads to the various buildings. Because of the danger of fire, smoking is only permitted in the parking areas. The quarter-mile path to the mill is over rough ground. For a complete visit, including the mill complex, allow from one to two hours.

Macon, Georgia, to Tifton, Georgia

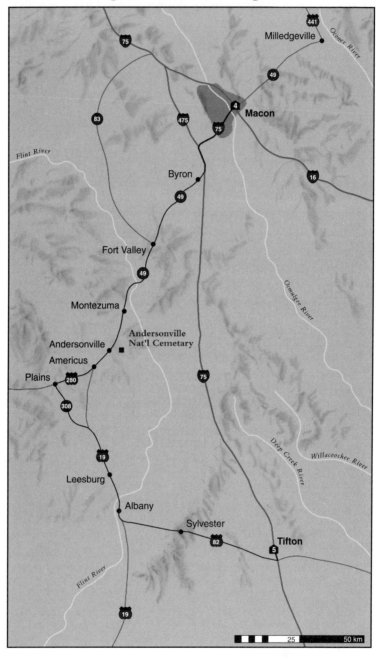

△ Day Five

MACON TO TIFTON, GEORGIA (142 miles)

A fall zone location alone did not guarantee that a town would become a successful entrepot, but Macon's river, the Ocmulgee, proved to be navigable, and a major inland port developed on the river. Prior to the introduction of steamboats in 1829, Macon was served by a flotilla of poleboats that connected the town with coastal Darien. The first steamboat pushed up the Ocmulgee in 1829 and in 1833 regular service to the coast was begun.

By 1837 seven steamboats provided Macon with regular service to the coast and some sixty towboats were in use. Lumber and cotton were the chief cargoes shipped through Macon in this period. Thanks to Savannah's desire to tap the cotton production of the Piedmont's plantations and farms, Macon became an important rail hub by 1840.

Macon to Marshallville, I-75, Ga 49, 32 miles

Driving south from Macon the traveler leaves the fall zone and enters the Coastal Plain. To the traveler the transition from Piedmont landscape to Coastal Plain landscape is not sharp or abrupt. Seen from space, however, contrasting land-cover and land-use patterns are clear and unequivocal. Color-enhanced satellite imagery shows the wide, swampy floodplains of Coastal Plain rivers as forming strikingly broad, dark ribbons extending to the coast. The

Stream and Valley Characteristics, Ocmulgee River

Stream Gradient

Valley Length - Valley Width

Source: Compiled from Topographic Maps and Field Data

broadening of the Ocmulgee River's valley downstream from Macon is among the most dramatic. Streams become more sluggish and serpentine when they leave the fall zone and flow across the Coastal Plain.

Paralleling the fall zone across Georgia is a highly dissected belt with elevations in the range from 250 to 750 feet. This is known locally as the fall-line hills. In this belt Georgia's chief mineral resource, kaolin, is mined. Georgia dominates U.S. production of kaolin and associated clay products. The traveler should be alert to evidence of kaolin production in this area.

Byron, in Peach County (14 miles south on I-75), was incorporated in 1874. It owes its name to the literary leanings of its founder, a Mrs. Richardson, who was an admirer of the great English romantic poet. In more recent years Byron, Georgia, gained notoriety in 1970 as the site of the state's largest ever pop-music festival and "skinny dip." Given his reputation for love of the unconventional, the original Byron was doubtlessly pleased by the distinction of his namesake.

Fort Valley (Ga 49, 11 miles) is the seat of Peach County and home of Fort Valley State College. The town began in 1820 as an Indian trading post and there is no record of a fort of any sort ever being in this area. According to local lore the site was originally known as Fox Valley. However, in 1825, the name was misread on a poorly written post office application. Once postal authorities began using the name Fort Valley rather than Fox Valley it stuck. In April this area of Georgia is at its most beautiful thanks to the peach blossoms that brighten so much of the landscape. The traveler will have no trouble identifying peach and pecan orchards. Peach trees are generally much smaller than the lofty pecans. The choice of the county name is clear evidence that the cultivation of peaches dominated the lives of most citizens in the area by the 1920s when the county was created.

The Massee Lane Garden, home of the American Camellia Society, can't be passed by. Begin with the fifteen-minute slide presentation. Brick pathways lead the visitor through nine acres of camellias under towering southern pines. The best season for enjoying the garden is from November through March when it is open daily. Admission is charged for visitors over twelve.

Marshallville is the oldest town in Macon County. Local lore identifies the first settlers as South Carolinians who began farming here in the 1820s. The town was named in honor of a much-loved Methodist preacher, Rev. John Marshall.

✓ Marshallville was the home of Samuel H. Rumph, originator of the Elberta peach. He honored his wife by borrowing her name for a fruit which proved to be ideally suited for canning and became known around the world. Rumph's home is on the right side of Main Street as the traveler follows Ga 49 through town. Built in 1904, the house reflects the wealth made in the fruit industry that developed in the aftermath of the Civil War.

Down the local road on the right is a bridge crossing the Flint River. Until very recently it was the location of Georgia's last free ferry. In earlier times ferries and fords were frequently encountered as travelers crossed the South's innumerable streams and rivers.

Marshallville to Montezuma and Oglethorpe, Ga 49, 16 miles

These sister cities are located respectively on the east and west side of the Flint River. Oglethorpe, named for Georgia's colonial founder, was incorporated in 1845, and is the older by some five years. Montezuma, named by returning Mexican War veterans, owes its existence to the spread of railroads in the antebellum period. It began to develop when the Central of Georgia came through this area. The coming of the railroad also caused the rapid growth of Oglethorpe, which soon boasted of being the "metropolis" of southwest Georgia. Boosterism succeeded in having the government of Macon County moved to Oglethorpe in 1854 but failed in similar attempts to have the state government moved there from Milledgeville.

The accessibility of the railroad brought great commercial gain to the towns and cities it served and to landowners like E. G.

Cabaniss. He ran an advertisement in the *Macon Journal and Messenger* on October 24, 1849, that concluded:

> It is now certain that the railroad will be completed to that point in time for the crop of 1850. The grading is progressing rapidly and the iron for the road as far as Oglethorpe is contracted for, to be delivered in Savannah next January.
> The attention of capitalists and of all who may wish to share in the business which will spring up in a place where at least 70,000 bales of cotton will be sold annually is respectfully invited to the sale of lots above specified. An inspection of the map will show that it is the point where the business of the great cotton-growing section of Georgia will concentrate.

Andersonville, 7 miles south on Ga 49, is in Sumter County and was named after the first recorded settler to make his home in the area. First known simply as Anderson or Anderson Station, the town owes its fame to its being chosen in 1863 as the site for what became the largest Civil War prison camp in the Confederate States of America. Camp Sumter, as it was officially named, was built in early 1864 near the terminus of the Southwestern Railroad. The modern town is adjacent to the Andersonville National Historic Site and the Andersonville National Cemetery.

The traveler should begin his or her visit with a stop at the welcome station and museum located in the nineteenth-century railroad depot. Hostesses are on hand seven days a week to welcome and assist visitors. Restaurants, antique shops, and bed-and-breakfast accommodations are listed here. A town RV park offers complete hookups. Architecture buffs should not miss a visit to the log church Pennington St. James, which was designed by the same architects who designed the Cathedral of St. John the Divine in New York City.

Andersonville National Historic Site

Unique in the National Park System of the United States, this park serves as a memorial to all Americans who have ever been held as

A NOTE ON MALARIA IN THE PLANTATION SOUTH

"He ain't sick; he's only got ager"—such was the ubiquity of malaria over the humid southeastern part of the United States. Geographer Melinda S. Meade has pointed out that plantations provided a concentration of population in rural surroundings, both conducive to the efficient transmission of malaria, the "ager" or ague of Southern vernacular speech. In her essay on "The Rise and Demise of Malaria" (*Southeastern Geographer,* November 1980), she quoted from mid-nineteenth-century observations published in Frederick Law Olmsted's *A Journey in the Seaboard Slave States:*

> Even those few who have been born in the region, and have grown up subject to the malaria, are generally weakly and short-lived. The negroes do not enjoy as good health on rice plantations as elsewhere; and the greater difficulty with which their lives are preserved, through infancy especially, shows that the subtle poison of the miasma is not innocuous to them; but Mr. X boasts a steady increase of his negro stock of five per cent per annum, which is better than is averaged on the plantations of the interior. As to the degree of danger to others, "I would as soon stand fifty feet from the best Kentucky rifleman and be shot at by the hour, as to spend a night on my plantation in summer," a Charleston gentleman said to me.

The modern traveler viewing the relict features of the old Plantation South will certainly find mosquitoes and other insects annoying. What we should remember is that indigenous malaria was a scourge throughout the history of plantation agriculture in the South. Not until the 1950s was the disease, for all intents and purposes, eradicated in the United States.

prisoners of war. Begin a tour at the visitors center where Park Service personnel provide information and direct you to the twelve-minute introductory slide show and relief map model showing historic and modern features of the park area.

During the American Civil War, no universally accepted convention governed the treatment of prisoners of war. Generally speaking the South's policy favored prisoner exchange to help refill their often undermanned ranks. The North, on the other hand, rarely initiated prisoner exchanges and so their surrendered forces placed an extra burden on the Southerners. The Confederates deemed Camp Sumter necessary to house the large number of prisoners being moved away from camps in and around Richmond, Virginia, in 1863. During the fourteen months of its use as a military prison more than 45,000 Union soldiers saw confinement here. Over a quarter of them died from disease, malnutrition, and exposure in the prison pen. A fifteen-foot-high stockade of pine logs surrounded a sixteen-and-a-half–acre area.

As the Confederacy neared its end, conditions at Andersonville went from bad to horrific. In his diary Michigan cavalryman John Ransom wrote of fellow prisoners "with sunken eyes, blackened countenances from pitch pine smoke, rags, and disease, the men look sickening. The air reeks with nastiness." When the war ended the camp's commander, Captain Henry Wirz, was arrested and, in the judgement of many, made a scapegoat. Wirz was tried and sentenced to be hanged by a military tribunal amidst a near hysterical public outcry against the atrocities of war. He was executed on November 10, 1865. The United Daughters of the Confederacy erected a monument in Wirz's honor that stands in the town of Andersonville. It helps the visitor capture some of the depth of the division that the Civil War created in the minds of our forebears.

Two miles south of Andersonville on Ga 49 is the entrance to historic *Trebor Plantation,* first established in 1833 by Robert James Hodges, who built the Greek Revival house that is the centerpiece of the plantation. Many of the plantation outbuildings date from the era of the main house, but others were added later as the farming emphasis shifted away from cotton. An admission charge is required to visit Trebor Plantation.

Andersonville to Americus, Ga 49, 9 miles

Americus, founded in 1832, was made the county seat of Sumter County in 1852. Americus derives its name from America, the name applied to the western continents by European geographers and cartographers in the early sixteenth century. Georgia Southwestern College, located here, became part of the University System of Georgia in 1932. Americus, the masculine form of America, was chosen in a drawing held to decide the town's name. One of the most impressive Victorian buildings in the South houses the historic Windsor Hotel. Recently restored and refurbished, the hotel has a round tower and upper veranda overlooking downtown Americus.

In the early years, Americus residents had to travel to the now-extinct town of Danville, 16 miles to the east, for many of their needs. Danville, at the head of steamboat navigation on the Flint, flourished and rivaled Americus until the coming of the railroad. In 1891 an old resident wrote that when his family arrived in 1849 "the principal town in the county was Danville. Circuses would skip Americus to go to Danville." By becoming the county seat and attracting a railroad linkage Americus continued to grow and flourish while Danville went into decline. Today nothing remains on the quiet bluff overlooking the Flint River where the bustling river port of Danville once stood.

Depart Americus going west on US 280 and parallel the railroad 11 miles to Plains in western Sumter County.

Plains to Albany, Ga 308, U.S.19/Ga 3, 31 miles

Plains is best known as the hometown of James Earl "Jimmy" Carter, Georgia's seventy-sixth governor and the thirty-ninth president of the United States. President Carter was born here on October 1, 1924, and attended the local public schools. In 1943 Carter entered the U.S. Naval Academy in Annapolis, Maryland, where he graduated in 1946. The death of his father later caused him to abandon his naval career and return to Plains to help manage the family seed and farm-supply business.

Mention of Plains and Jimmy Carter usually suggests Georgia's chief crop—peanuts (or groundnuts). As cotton began to go into decline, peanuts became increasingly important across the old cotton belt. By 1943 Georgia was the leading peanut-producing state, an honor it still holds. Mr. Peanut quietly nudged King Cotton off his throne in the post–World War I decades. Plains is in the heart of Mr. Peanut's realm.

"Ground peas" is a more accurate description of the crop in the lush green fields that cover much of Georgia's inner Coastal Plain. Peanuts grow underground and are not nuts at all. During the Civil War they were known as "goober" peas, from an African word for groundnuts. Peanuts are planted in April and are usually ready for harvest in early autumn. When the shell-like pods clinging vinelike to the lower branches of the plant are ripe, the plants are turned rootside-up and left to dry in the sun for a few days. When dry, a peanut combine separates the nuts from the vine. The nuts are then shelled, cleaned, dried, and processed for marketing or storage. More than half of Georgia's edible peanuts go into peanut butter.

The beautifully landscaped park in the center of Plains was donated by the people of Kaohsiung, Taiwan, in 1978. Plains is Kaohsiung's "twin" city.

Depart Plains via Ga 308 heading southeast into Lee County. Turn south on U.S. 19/Ga 3 to Leesburg.

Leesburg is the county seat of Lee County. The county was created in 1825 from land acquired from the Creek Indian Nation and was named in honor of the Revolutionary War firebrand General Richard Henry "Lighthorse Harry" Lee of Virginia. Lee became a hero in Georgia by taking Augusta from the British in 1781. Lee County is in the heart of Georgia's farm belt: nearly three-quarters of the land in the county is in farms, while the state average is only a third.

Chehaw Park is located just a few miles north of Albany, Georgia. In addition to RV and camping facilities, it features boat access to Lake Worth on the Flint River. A wild-animal park and petting zoo make Chehaw Park particularly popular with children. The National Indian Festival is held here over the third weekend in May every year. The festival is in the spirit of Black Elk who prophesied that a time would come when "the sons and daughters of our oppressors will return to us and say, 'Teach us so that we might survive, for we have almost ruined the earth.'"

Albany to Tifton, U.S. 82, 34 miles

Although named for the capital of New York State, local pronunciation makes Albany, Georgia, sound like "All ben'ny" or "Al bain'y." Albany's history is closely tied to the Flint River which flows through the town on its way to join the Gulf of Mexico. The town's founders in 1836 felt that their city, near the head of navigation on the Flint River, would prosper in the way that Albany, New York, had near the head of Hudson River navigation. But hopes for regular steamboat schedules were frequently frustrated by unreliable water levels and channel obstructions that characterized the lower Flint-Apalachicola route.

By the time the Civil War broke out Albany and Thomasville marked the railheads in the southwest corner of Georgia and the future of the former river port was assured. In 1913 the Union Passenger Station was built to provide a link for the seven railroads then providing Albany with as many as thirty-five trains daily. The station now serves as the home of the *Thronateeska Museum of History and Science*. Of special interest is the museum's King Cotton exhibit. Albany grew by incorporating land formerly embraced in several nearby cotton plantations. The *Smith House* on Flint Avenue is regarded to be the first brick home in Albany and is an excellent example of late antebellum construction. Built before the railroad was completed, the house's bricks were shipped by wagon from Macon. An unusual feature for its period is the indoor water system supplied by cisterns in the attic.

Depart Albany heading east on US 82 toward Tifton. The highway parallels the Seaboard Coast Line railroad over the almost imperceptible divide that separates the Atlantic and Gulf of Mexico drainage systems in this part of Georgia.

As the small settlement named *Acree* is passed the traveler crosses from Dougherty to Worth County. A close scrunity of a map will reveal that this boundary, in common with several others in western Georgia, is crenelated. The otherwise straight-line boundary (north-south in this case) is marked by rectangular offsets reminiscent of castle battlements. In an article in *Essays on the Human Geography of the Southeastern United States* (West Georgia College, Carrollton, Georgia, 1977), Florida geographers Burke C. Vanderhill and Frank A. Unger pointed out that 70 of the 159 counties in the State of Georgia exhibit what can be termed crenelated boundaries. Crenelation is only found in the Land Lottery region of Georgia and adjoining Alabama counties—nowhere else in the United States.

Rather than the product of political gerrymandering, Georgia's crenelated county boundaries are a memorial to an age when individual freedom was better represented by government than is the general rule today. Until 1879 boundary crenelations usually resulted from the favorable actions by the state legislature in response to the requests of private landowners to be included within the limits of an adjoining county. In all, some 531 county boundary changes resulting in crenelation were passed by the Georgia legislature in the half-century beginning in 1828. In 1879 the responsibility for boundary changes were transferred to the affected county governments. Such things as tax differentials, personal allegiance to a particular town or community, ease of access to the county seat, desire to stand as a candidate for elected office, and desire to have a residence in the same county as business interests were factors contributing toward boundary crenelation.

Sylvester, the county seat of Worth County, was originally known as Isabella Station; the present name emphasizes the importance of forests to this part of Georgia. Most authorities agree the name was adopted in 1894, and was derived from the Latin *silva,* meaning "wood," and *vester,* meaning "your." The name Isabella still lives on in the hamlet on Ga 313 one mile north of Sylvester. Today,

94   THE PLANTATION SOUTH

Sylvester presents itself as the "Peanut Capital of the World" and hosts an annual peanut festival during the second weekend in October. Contestants from throughout the state compete for the title "Georgia Peanut Queen" in one of the festival highlights.

Ty Ty in Tift County was incorporated as a town in 1883 and took its name from a nearby creek. According to one local legend the first postmaster of Ty Ty suggested the name because of the importance of the railroad tie industry in the area at that time. However, geographer Roland Harper, writing in the *Albany Herald,* May 23, 1953, suggested that the name derived from the vernacular expression "tighteye" meaning a thickly vegetated place where it was hard to see. Still others argue that this fascinating name is rooted in some Indian language, perhaps Creek or Uchee.

Tifton, the county seat of Tift County, was first settled in 1872. The county was named for Colonel Nelson Tift (1810–1891) who helped found Albany. Once Georgia's leading tobacco market, Tifton now boasts of being "The Tomato Plant Capital." Located a short distance north of the tree-shaded streets of the town of Tifton and campus of Abraham Baldwin Agricultural College is the state's museum of agriculture, Georgia Agrirama.

Georgia Agrirama is a living history museum representing the state's agricultural life in the years between the Civil War and the twentieth century. Here, for a small admission charge, the traveler can step back into the nineteenth century and enjoy a close-up look at Georgia's rich rural heritage. The daily round of typical farm activities are carried out in authentic settings by a staff of well-informed and appropriately costumed professional interpreters. Contact Agrirama for a schedule of special seasonal events to better plan a visit to this uniquely entertaining and educational museum.

△ Day Six

TIFTON TO JEKYLL ISLAND, GEORGIA (131 miles)

The Pine Barrens

The "pine barrens" of Georgia and South Carolina are portions of the vast pine forest that swept in a great crescent from the lower Chesapeake Bay to the Mississippi River. This truly grand belt of trees form a forest a thousand miles in length and scores of miles wide. It dominated much of what is known today as the Atlantic-Gulf Coastal Plain, that area of relatively flat, sandy lands stretching from the rolling sand-clay hills of the fall zone to the coastal estuaries and tidal marshes. The pine barrens are home to some eight or ten species of what are popularly known as yellow or pitch pines. The most important of these are the long-leaf *Pinus palustris*, short-leaf *Pinus echinata*, loblolly *Pinus taeda*, and slash pine *Pinus ellioti*. Deep taproots permit the pines to reach far into the sun-blasted sandy soils, but something more is needed to explain their near total dominance over such a vast range.

That explanation is to be found in the realization that fire, along with soils and rainfall, is an essential element in the ecology of the Southern pine forest. The work of E. V. Komarek and his associates at the Tall Timbers Fire Ecology Center has shown that without fire the pine forest gives way to a hardwood-dominated association. Climatically the forest region presents optimum conditions for "natural" or lightning-induced fires. American Indians

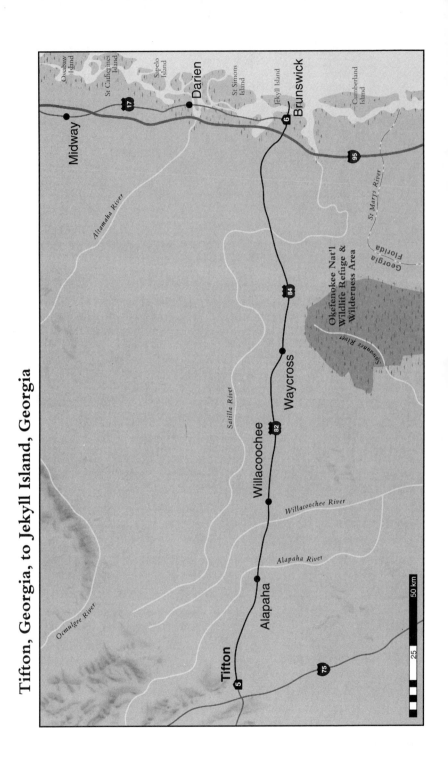

Stand of naval stores pines probably from along the Altamaha River, after 1930.

Man preparing to harvest turpentine in the pine barrens.

Day Six 99

Men loading logs to truck them out of the forest.

Rafters moving logs down the Altamaha River, probably before World War I.

and pioneer European settlers employed periodic burning as a landscape management tool.

The veteran English traveler Captain Basil Hall was overwhelmed by the immensity of the virtually trackless forest of pines. He found travel through the forest south of the Altamaha River "like navigating by means of the stars over the trackless ocean!" "For five hundred miles," he continued, "we travelled in different paths of the South, over a country of this description, almost everywhere consisting of sand, feebly held together by a short wiry grass, shaded by the endless forest."

Anecdotal observations stress the singularity of the pine-dominated forests; the reality is more complex. The seemingly endless and unbroken expanses of pine were often interspersed with other forms of vegetation. Inland from the coast there were occasional gentle elevations known to explorers and pioneers as "hummocks," "homocs," "hommocks," "hammacks," or "hammocks." On such spots hardwoods grew. It was soon learned that their presence indicated a soil change for the better, and hammocks were sought out by early settlers in the pine barrens.

Here and there the somber pines gave way to open glades or natural meadows that the pioneers knew as "savannahs." Often these places were slightly lower and relatively moist. The lush native grasses that grew in the savannahs and attracted early herdsmen, the first settlers in the pine barrens. Their cattle fed on the rank wiregrass and cane that grew in glades and savannahs. Extensive "brakes" of reed cane, described as highly desirable by the early herders, then existed along many streams coursing through the barrens. Overgrazing, fire, and drainage of bottomlands all contributed to the disappearance of the areas known as "reedy branches" and "caney flats."

Remaining on the present map and landscape are frequent reminders of the importance of the period of stock grazing in the pine barrens. Numerous stream names include terms such as "caney" or "reedy" that hark back to the old cattle days. Also, creek names containing the word "moss" or "mossy" were connected with stock-raising; they refer to a vegetation known as "moss" or "salt grass" that grew on rocks in the streams. Cattle were particularly fond of this vegetation and would wade in the water to reach it. Place-names found on old maps that referred to aspects of the cattle industry, such as cowpens, have nearly all fallen out of use. Cowpens Battlefield in South Carolina and Cowpen Mountain in Georgia are notable exceptions. Cowfords on larger rivers were also significant places. Probably the best known in colonial times was on the St. Johns River in northern Florida where Jacksonville was first known as "Cowford."

When development finally arrived in the pine barrens it came with remarkable speed. The magic of the railroad unlocked the potentials of the great pinelands, opening this vast reservoir. The railroad gradually threaded the pine barrens after the Civil War; and by the 1890s, the region began to boom. By the turn of the century it was the leading lumbering belt of the nation and it became the chief naval-stores-producing region of the world. Towns sprang up overnight.

In this rush the expression "pine barrens" was quickly forgotten. One Floridian said the name was all wrong anyway, because he thought the term was a corruption of "pine bearing," an expression which was more descriptive of the true potentialities of the area.

Fortunately for the region, pines reproduce easily and grow fast if given an opportunity. With this great advantage the pinelands were able to retain a place as an important lumber and turpentine-producing section. In recent decades a big new industry, the pulp and paper business, has offered further outlets for forest resources in the form of pulpwood.

Today the area formerly known as the great pine barrens is a prosperous region sprinkled with thriving communities and pretty towns. Few people living there nowadays have heard the name, and would be reluctant to believe that it was ever applied to their locality.

Tifton to Waycross, U.S. 82, 71 miles

Leaving Tifton and heading east on US 82 the traveler soon passes through the village of *Enigma* in Berrien County. Curiosity makes one wonder how the town got its name. Sorry, but it is still an enigma. Berrien County, on the other hand, was named for John MacPherson Berrien (1781–1856), who served as the attorney general of the United States in the administration of President Andrew Jackson and was a leader in Georgia's nullification movement in the early 1830s.

Alapaha was incorporated in 1881 and named for the nearby Alapaha River that flows into the Suwannee River in Florida. Old timers in this part of Georgia pronounce the name "Loppy-haw." There is a difference of opinion as to whether the word comes from the Creek or Timucua Indian languages.

Willacoochee is located on Willacoochee River at the western edge of Atkinson County. It was incorporated in 1889 and marks the crossing point of the Central of Georgia and Seaboard Coast Line Railroads. A traveler interested in the history of the naval stores industry will want to visit *McCranie's Turpentine Still* just outside the Willacoochee city limits. It is the best remaining example of a still utilizing the fire-burning distillation process, a method once employed throughout the pine barrens of the Southeast. There is no permanent staff at the still, so the visitor should make arrangements in advance by contacting the city clerk.

Pearson, the county seat of Atkinson County, was incorporated in 1890. Pearson takes its name from a prominent local citizen who had served with distinction in the Indian War of 1838. The first residence in Pearson was built in 1873 by S. J. Henderson and is still standing. It is the *Minnie F. Corbitt Memorial Museum* and can be visited on request.

The county seat of Ware County, *Waycross* was incorporated in 1874 as "Way Cross." Local tradition maintains that the name was chosen by early community leaders because so many roads crossed here. Ware County, encompassing some 912 square miles, is the largest in area of any county in Georgia. It takes its name from Virginia-born Nichols Ware (1769–1824), a former mayor of Augusta who served in the U.S. Senate from 1821 until his death. Waycross is the traveler's gateway to world-famed Okefenokee Swamp.

Turn left on North Augusta Avenue to *Okefenokee Heritage Center* and *Southern Forest World*. Southern Forest World is a free educational exhibit center devoted to the development of forestry in the thirteen southern states. A small charge admits the traveler to the adjoining Okefenokee Heritage Center where well-designed exhibits tell the fascinating story of the people living in and around Okefenokee Swamp. After a visit drive south through the downtown area and follow Swamp Road to "Obediah's Okefenok."

Obediah's Okefenok is the restored homestead of Obediah Barber who was born in 1825 and died in 1909. He and his father Israel were among the first whites to settle on the northern edge of Okefenokee Swamp. The one-story cabin Obediah built and lived in has wooden-pegged walls and log-puncheon floors. Other buildings and exhibits give a rich view of the lifestyle common to this unique environment from initial settlement to the early 1900s. A small admission is charged. Return to Waycross and follow US 1 south to Okefenokee Swamp Park.

Okefenokee Swamp Park

Okefenokee Swamp Park gives travelers access to this always awesome and sometimes mysterious swamp. In his famous book *Travels Through North & South Carolina, Georgia, East & West Florida* (1791), naturalist William Bartram described Okefenokee thus: "The river St. Mary has its source from a vast lake or marsh, called Ouaquaphenogaw, which lies between Flint and Oakmulge rivers, and occupies a space of near three hundred miles in circuit. This vast accumulation of waters, in the wet season, appears as a lake, and contains some large islands or knolls, of rich high land; one of the present generation of Creeks represent [it] to be a most blissful spot of the earth."

The British surveyor general during the decade before the American Revolution, William Gerard DeBrahm, wrote "this swamp and its contiguous lands are said to form a Country of exceeding rich soil. The Jealousy of the Indians has not as yet permitted me to make a Survey of it, for this whole Swamp and its contiguous Country is contained within the Hunting Ground reserved by the Indians." The

104 THE PLANTATION SOUTH

Hunters in the Okefenokee Swamp.

archaeological evidence from Okefenokee indicates that the Creek Indians and their offshoots the Seminoles hunted in the swamp but didn't reside there except as a refuge during periods of stress.

After examining the natural flora and fauna of Okefenokee, visitors should drive north to US 84 and continue heading east toward the coast.

Waycross to Jekyll Island, U.S. 84, Ga 50, 60 miles

Nahunta is the county seat of Brantley County. The name is thought to be Iroquoian, which seems strange in this area of former Creek Indian domination. Its meaning is believed to be "tall trees" and was a tribute to the pine barren forests. Local experts say that the name was suggested by a turpentiner who had moved in from Nahunta, North Carolina. Brantley County was created in 1920 and named for a prominent resident.

Turn onto Ga 50 and continue to Jekyll Island.

Now tourists are more frequent visitors than hunters, and bridges have been built through the swamp to accommodate them. Photograph by Lynda Sterling.

Jekyll Island and St. Simons Island, Georgia

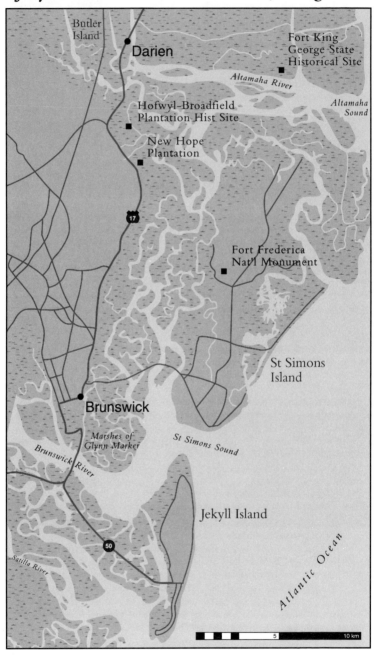

◸ *Day Seven*

JEKYLL ISLAND AND VICINITY

The early-rising traveler is in for a treat on Jekyll Island; as the sun appears over the Atlantic Ocean the term "Golden Isles" cannot help but come to mind. Burnette Vanstory in her 1956 book, *Georgia's Land of the Golden Isles,* writes that early treasure seekers may have thought of coastal island beaches as golden as they sought treasure washed up from foundered Spanish galleons coursing north on the Gulf Stream offshore. The name "Golden Isles," however, was introduced to the world in 1717 by Sir Robert Montgomery, a Scot with a charter to found a colony in the "debatable land" lying between English Carolina and Spanish Florida. Montgomery extolled the beauties and wonders of coastal Georgia in a tract published under the flamboyant title "A Discource Concerning the design'd Establishment of a New Colony To The South of Carolina In The Most Delightful Country of the Universe." Despite this major promotional effort his charter lapsed before he could start a settlement.

Jekyll Island is one of the smaller of the range of barrier islands that flank the coasts of Georgia and South Carolina. The islands came under Spanish influence in the sixteenth century in the form of missions and presidios sent out from St. Augustine to convert the region's Indian populations. By the mid-seventeenth century, Spanish Franciscan missionaries served some 25,000 Indians in thirty-eight missions with as many as seventy friars. At the same time only twenty-six friars were active in the Southwest.

As archaeologist David Hurst Thomas frequently emphasizes, "the southeastern mission system was founded earlier, involved more people, and lasted longer than the southwestern system which is so familiar to Americans." Thomas has done much to correct public impressions through his excavations of Santa Catalina de Guale on St. Catherines Island, Georgia's oldest confirmed European settlement. The establishment of a colony in South Carolina by the English in 1670 eventually led to the end of the Spanish mission system. After an attack on Santa Catalina in 1680, the Spanish abandoned the mission; by 1686, all of the missions north of Florida were gone. The area now occupied by Georgia soon became a "debatable land" contested for by Europe's leading imperial powers.

After 1733, English colonization in Georgia brought a plantation agricultural system that extended to the barrier islands. It was first based on rice and indigo and later on the cultivation of Sea Island cotton. Entire islands became the domain of groups of wealthy, slave-owning planters. The last shipload of slaves from Africa to America came to Jekyll Island in 1858. They were brought aboard the sailing ship *Wanderer,* built in Maine for the illegal slave-running trade. By the time federal authorities got wind of the illegal landing, most of the blacks had already been purchased. The ship's officers and owners were brought to trial in Savannah but no convictions were forthcoming. A large riveted metal "mess kettle," reportedly used for preparing the slaves' food on the *Wanderer,* is on exhibit at the *Jekyll Island Orientation Center Museum.* The Civil War brought devastation and ruin to Jekyll's slave-dependent society as the vulnerable islands proved impossible to defend against Union raiders and invaders.

The Jekyll Island Club Hotel is a reminder of another occupation of Jekyll Island. In 1886 the island was purchased by a small group of America's richest families—including Rockefellers, Morgans, and Pulitzers—who formed the Jekyll Island Club. The club was part of a trend that saw many of the barrier islands purchased by wealthy Americans eager to enjoy their natural beauty, beaches, and mild winters. This wave of conspicuous consumption had an unforseen dividend in that it saved many of the islands from the piecemeal development which has diminished the attractiveness of so much of the Atlantic seaboard farther north. Preserved by the heirs of the nineteenth-century robber barons, many islands were little altered from their natural states until they were acquired by government and preservation trusts. After its acquisition by the state of Georgia in 1947, Jekyll Island, playground of the wealthy, was converted to a playground for the common people. The Jekyll Island Authority, which administers the island, boasts that it offers a wide range of recreational and educational opportunities that are easily affordable for family vacations.

Jekyll Island was named by Georgia's founder James Edward Oglethorpe in honor of Sir Joseph Jekyll, who helped back the establishment of the colony. Oglethorpe was exceeding his authority, however, because the original geographical limits of Georgia were the Savannah River on the north and the Altamaha River on the south. Jekyll Island is several miles south of the Altamaha.

The *Marshes of Glynn* marker and overlook park honor Georgia poet Sidney Lanier, who extolled the state's coastal marshlands long before modern-day ecologists revealed the value of their unique ecological treasure. Lanier wrote in "The Marshes of Glynn":

And what if behind me to westward the wall of the woods stands high?
The world lies east: how ample, the marsh and the sea and the sky!
A league and a league of marsh grass, waist high, broad in the blade,
Green, and all of a height, and unflecked with a light or a shade,
Stretch leisurely off, in a pleasant plain,
To the terminal blue of the main.

Brunswick

Brunswick is the county seat of Glynn County, Georgia's second busiest seaport, and home to an active fishing fleet. Founded and laid out in 1771, Brunswick is regarded as the last town established by the British in what is now the United States. Surveyor George Mackintosh surveyed a Savannah-like pattern of streets and open squares "at a place called Carr's Old Fields" on a bluff overlooking the Turtle River. To ensure public access to the river's navigable channel, the royal council stipulated "that the land under the Bluff of the said town [Brunswick] Opposite to each Street fronting the River and the full Wedth [width] of each Street respecting be reserved for Publick Landing places" (Colonial Records of Georgia XI, 386). The name Brunswick honors England's monarch, King George III, whose family was linked to the house of Brunswick.

In spite of being laid out on a bluff, Brunswick is only fourteen feet above sea level—the lowest elevation of any city in Georgia. On the way to the riverside harbor, the square at Reynolds and G streets is especially noteworthy. It contains the *Glynn County Courthouse* in a parklike setting of moss-draped live oaks. The Brunswick–Golden Isles Chamber of Commerce is located in historic *Dart–Brown House,* which dates from the 1880s.

Brunswick hosts a colorful Blessing of the Fleet ceremony and a seafood festival each May to mark the beginning of the fishing season. Good food, a parade, and brightly decorated boats make this a popular event.

Heading north from Brunswick on US 17, on the left is Glynn Jetport, formerly a Naval Air Station. This explains its enormous size. To the traveler's right, and due east across a three- to five-mile-wide belt of marsh and tidal creeks, is St. Simon Island.

The Rice Coast

Opinion differs as to when rice cultivation began in the American colonies. South Carolinians were probably cultivating some rice even before a legendary rice-laden brig from Madagascar took refuge in Charleston harbor in the early 1690s. However, local tradition, albeit discredited by most experts, holds that a bag of Madagascar rice given to Landgrave Thomas Smith provided the seed from which the Carolina rice culture eventually sprang. From

Tidewater Rice Area

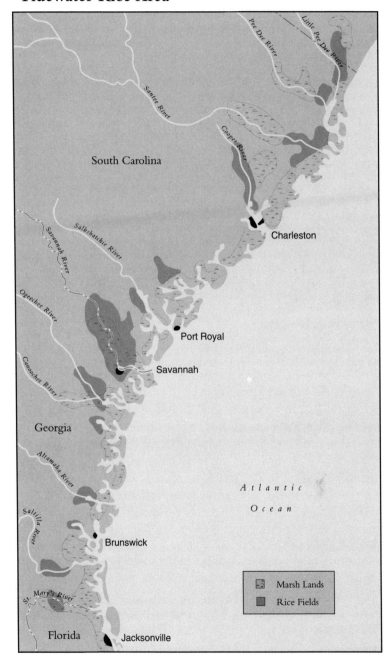

Coastal Plain swamplands.

wherever the first seed rice came, by the 1730s rice cultivation was well established in South Carolina. Within a year or two of Georgia's founding, rice was being grown experimentally in Savannah. The greatest impediment to the spread of rice cultivation into Georgia was the prohibition against slavery that was part of the Georgia original colonial plan. The prevailing opinion was that only the labor of Negro slaves could successfully bring land into rice cultivation.

The unavailability of slave labor plagued all large-scale attempts to develop Trusteeship Georgia. The indentured servants shipped in were found to be inadequate for the grueling tasks involved in bringing a subtropical Coastal Plain wilderness under cultivation. Patrick Tailfer and other frustrated landowners brought this unpleasant fact to the attention of the trustees in a communication received in London in August 1735. The landowners wrote that although they desired nothing more than to "settle upon and improve our Land" they found it impossible to accomplish this end "without the Use of Negroes." Their white indentured servants, they explained, "not being used to so hot a Climate can't bear the Scorching Rays of the Sun in the Summer when they are at Work in the Woods, without falling into Distempers which render them useless for almost one half of the Year." Clothing and feeding slaves was shown to be far cheaper than maintaining indentured servants.

The evidence suggests that slaves were being employed surreptitiously on remote landholdings in Georgia by the 1740s. In 1748 one settler stated, "its well known to every one in the Colony that Negroes have been in and about Savannah for these several years [and] that the Magistrates knew and wink'd at it and that their constant Toast is 'the one thing needful' by which is meant Negroes" (Colonial Records of Georgia, 25, 295).

Ultimately pressure for the legalization of slavery in Georgia became irresistible and the antislavery policy was repealed by 1750. In less than two years the trustees surrendered their colonial charter, Georgia became a royal colony, and any vestige of the Trust's utopian land and labor policies was eliminated. By 1772, the royal surveyor reckoned that Georgia had a total of 13,000 slaves, "Of whom are at least 12,000 employed in the Plantations." He continued by detailing the requisites and costs necessary for a successful pioneer rice plantation in Georgia: £2,476/16/0 was his figure. Of this, £1,800 was for the purchase of forty working hands.

The first rice plantations were developed on swamplands in the broad river floodplains of the Coastal Plains. Deep layers of alluvium provided the fertile soils sought by the rice planters as they spread toward the interior. On his way from Midway Church to the Altamaha River in 1773, Philadelphia naturalist William Bartram wrote of the extensive swamps forming the headsprings of the Medway and Newport Rivers. He noted that "these swamps are daily clearing and improving into large fruitful rice plantations." He described how the rice stood "in the water almost from the time it is sown, until within a few days before it is reaped, when they draw off the water by sluices." The water impounded in the up-slope reservoir also drove the milling machinery for cleaning the rice.

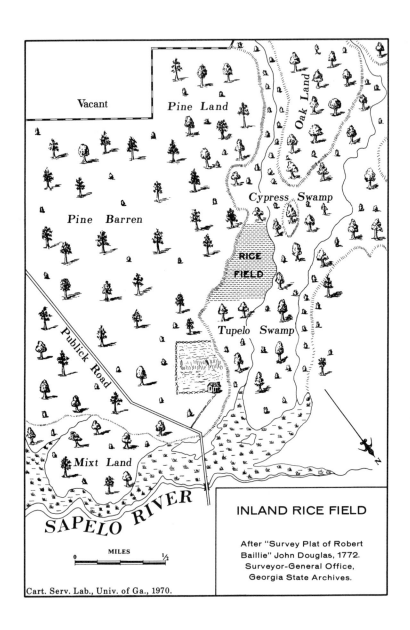

The shift in rice cultivation from upland swamp fields to estuarine fields flooded by the tides remains to be adequately documented by historical geographers. Georgia agriculture promoter Joseph Jones described the inland swamp rice fields as relics of a bygone age. Sam B. Hilliard, a geographer at Louisiana State University, is probably also correct in suggesting that in some instances a combination of both types of irrigation was used on those plantations where the tidal conditions allowed. He goes on to argue that, "despite the continued use of impounded fresh water for flooding, the advantages of tidewater rice culture led to its adoption by many planters and it soon emerged as the system of rice culture most characteristic of the area."

Tidewater rice culture occurred in the waters of estuaries from North Carolina to Florida where sufficient tidally-induced diurnal variation existed. On estuaries where proper conditions were found, prodigious inputs of labor by slaves wrought profound landscape modifications that remain dominant features in many coastal counties to the present day. Analysis of twentieth-century aerial photographs of relic rice fields show what Hilliard termed a "spotty distribution."

The first thing to recognize concerning this spotty distribution of tidal rice-producing areas is that in all cases the tidal rice fields are some miles removed from the ocean proper. A belt of outer islands, flanked by a parallel inner belt of tidal marshlands, occupy a broad zone between the sea and the rice-producing belt. In an ideal setting, the estuary dynamics took the form of a bottom-riding wedge of heavier salt water alternately pushing inland and withdrawing seaward. Any layering or mixing of salt and overlying fresh water created serious problems. The tidal rice producer depended upon the incoming tidal wedge to lift the surface layer of fresh water several feet at periods of high tide. The vertically lifted layer of fresh water could then be allowed to flow onto and flood low-lying, diked rice fields. When dry field conditions were desired the standing water was allowed to flow off the fields and into the river during low-tide periods. An appreciation of those hydraulics was written by the intrepid English traveler Captain Basil Hall, in his *Travels in North America, In the Years 1827 and 1828*. The

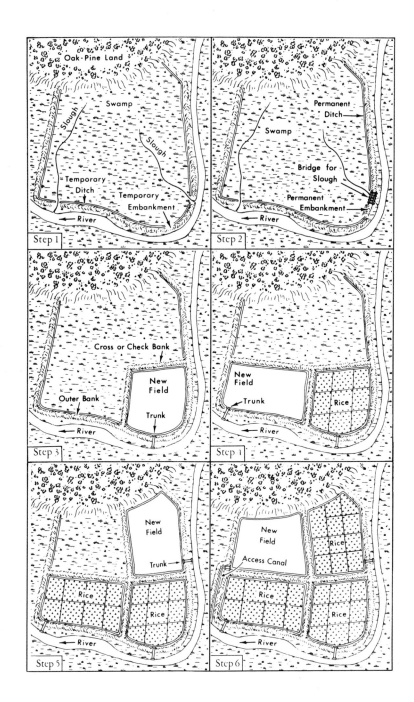

Detailed View of Irrigation Trunk and Gate

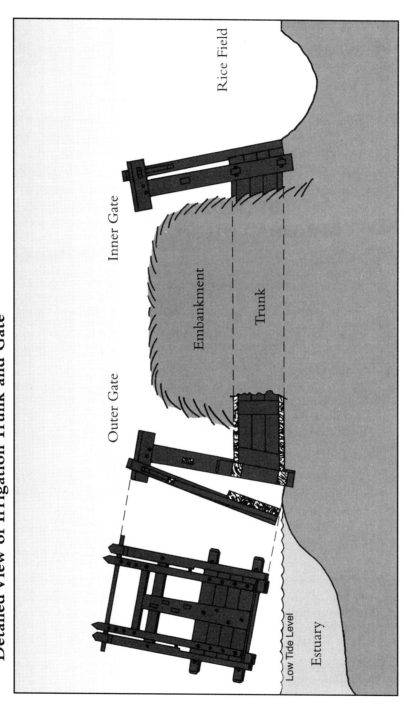

ever attentive Captain Hall also paid heed to the more industrial aspects of rice-coast life. He wrote in detail about milling rice and the advantages of shipping the rice before it has been husked:

> Rice with the husk on, or what is technically called Paddy—a word borrowed from India—will keep fresh and good for a much longer time than after it has undergone [processing]. . . . Besides which, prepared rice is apt to become dusty, either from exposure or from rubbing about in the carriage, on board ship, and in the warehouses on both sides of the Atlantic. These facts recently suggested to some enterprising capitalists to bring it to England in the shape of paddy, and there to detach the husk. This experiment has been completely successful, as I can testify from my own ample experience; for I have frequently, since my return, eaten rice managed in this way by Messrs. Lucas and Ewbank, of London, as fresh in taste and in appearance as any I met with in South Carolina.

At *Needwood Church,* the traveler sees to the right the transition zone on the south side of the Altamaha River above which tidal rice production was possible. Downstream conditions were excessively salty and rice was not grown.

NEW HOPE

New Hope Plantation was named and developed by a South Carolinian named Henry Laurens, a successful merchant and planter who served as president of the U.S. Congress during a critical period of the American Revolution. By the early 1770s he was one of the wealthiest men in the British colonies in America. Henry Laurens's landholdings included at least eight plantations in South Carolina and Georgia. His Georgia holdings included a 1763 grant from King George III for "a plantation or tract of land containing three thousand acres situate to the Southward of the Altamaha River bounded on the North East by the said River and Broughton Island and on all other sides by vacant land." In the years that followed its survey, Laurens developed the tract as a plantation he named "New Hope."

A study of the original eighteenth-century survey plats of the New Hope tract indicates that Laurens chose his land with care. A third of the long, mile-wide grant nearest the river was a tree-covered swampland subject to periodic freshwater flooding. Only the downstream extremity of the tract was marshland and too wet to support a native forest association of cypress, bay, water oak, gum, laurel, and tupelo trees. Away from this forested belt of swampland, the upland two-thirds of the land was covered with a sandy pine barrens broken by low patches identified as "Inland Swamps" by surveyors. The term "tide swamp" the surveyor included on the riverine portion of Laurens's 1763 survey plat was doubtlessly a welcome confirmation of the wealthy planter's hopes. It represented an eyewitness verification that his grant had the potential for tidal rice production once it had been cleared of trees and diked.

HOFWYL–BROADFIELD PLANTATION HISTORIC SITE

Upon arrival at Hofwyl–Broadfield the visitor should study the exhibits and attend the slide presentation at the visitors center. This state historic site makes the history of the Rice Coast accessible to visitors. In summer the traveler should wear protective clothing and use a repellent against insects before taking the path down to the relic rice fields. Use caution and avoid fire ants, snakes, or other wildlife that may be encountered along the way and the effort will be well rewarded.

From the visitors center proceed along the path toward the edge of the marsh where you will see several low blocks of tabby, oyster-shell cement. These ruins mark the site of what was probably a rice mill, built prior to the Civil War to process the winnowed rice. Because of high costs, few planters had mills; most sent their grain to mills in Savannah or Charleston.

Follow the path along the top of the rice-field dike. This freshwater marsh was probably once a virgin cypress swamp. Cleared by the hand labor of hundreds of African slaves, the junglelike swamp was leveled as flat as possible to insure proper drainage. Then the slaves constructed miles of ditches and dikes, also by hand. The dikes were used to hold water in the rice fields when required during the growing season. When the rice fields were flooded, foot travel was restricted to paths along the top of the dikes. Across the marsh notice the lines of small trees; these outline other dikes. Many of them are 170 years old.

To the left is a narrow channel down which the heavily laden rice flats were poled—bringing rice to the mill, carrying rice seed to be planted, or hauling earth to repair breaks in the dike system. The canals served a dual purpose: irrigation and transportation.

For most of the year wildlife abounds in the marsh. You may see a snake slithering off the dike into the water or fiddler crabs dancing on the bank. You may catch a glimpse of a raccoon, a wild hog, or one of the many large hawks which often circle the sky above the marsh. Butler Island Wildlife Management Area borders Hofwyl.

Although brackish water invades Hofwyl's marsh during storm-driven high tides, the marsh is basically freshwater. The plant life found here is the type found in freshwater environments. Salt water was deadly to rice and had to be kept out of the fields at all costs. If salt water flooded a field, the planter often had to wait years before he or she could plant rice there again.

From the rice-field dike, stroll to the front yard of the plantation house with its stately moss-laden live oaks (*Quercus virens*). From the high ground, look out over Hofwyl's 696 acres of marsh toward tree-lined Dent's Creek and the Altamaha River. Walk over to the flower garden with its brick-lined spring, which for years kept household dairy products chilled.

The plantation house is a modest and simple structure, probably built in the 1850s and modified over the years. In 1903, Dr. James Troup Dent screened the front porch, the windows, and the fireplace to prevent mosquito infestation, Prior to that time, plantation families generally abandoned the coast during the warmer months because they associated what we know as malaria with the "miasma" of the swamp, and not with the insect vector. The interior is furnished as it came to the state in 1973 at Miss Ophelia Dent's death.

As you leave Hofwyl, turn right on US 17 and proceed about a half mile before turning left on the secondary road leading to Sterling. Some feet down the road is an abandoned canal excavation.

The Brunswick-Altamaha Canal

Soon after the American Revolutionary War, state authorities and private investors began to promote internal improvement schemes aimed at facilitating the economic development of the new republic. Canals were deemed to be one of the most important undertakings in these development schemes.

In Georgia the Board of Public Works emphasized the importance of the Altamaha–Oconee–Ocumlgee river system in its 1826 report on waterway improvements. These rivers were providing crucial transportation links for an ever-increasing volume of cotton and other agricultural products shipped from the developing Piedmont. Darien, near the mouth of the Altamaha, was not effectively serving this commerce because of an inadequate natural harbor. As a result, speculators and investors in Savannah and Brunswick turned their attention to schemes designed to bring the Oconee–Ocmulgee–Altamaha trade to their warehouses and wharfs.

In 1826 the Brunswick Canal Company received a charter to build a canal or railroad to connect the Altamaha and Turtle rivers. After a long, drawn-out effort, excavation for a canal began in 1836. One of the greatest difficulties encountered was securing a reliable labor force. Irish workers were brought from Boston but they "kept the whole country in alarm by their drunken riots and vagrant habits." Soon the Irishmen were dismissed and four hundred slaves were brought in to replace them.

When Sir Charles Lyell, a geologist, visited nearby Hopeton Plantation in 1846, he wrote that nine miles of the canal had been completed at a cost of $900,000 but construction had ceased in 1839 because of a financial panic. In 1852 construction on the canal resumed. A mixed force of white laborers and five hundred rented black slaves were employed in completing the project. Public festivities to mark the twelve-mile canal's opening were held on June 1, 1854, almost fifteen years after construction began. However, the canal was not profitable and it was totally abandoned in 1860. Despite attempts at revival, the Brunswick-Altamaha Canal never became a viable artery of trade and commerce.

The last vestige of the human- and animal-dug canal stretches away from the crossing of Ga 99, in a ruler-straight line, south to the Turtle River near the modern port of Brunswick. Upon returning to US 17, turn left and proceed across the maze of Altamaha distributary channels and islands toward Darien on the river's north bank.

Butler Island

At one time the whole of the 1,500-acre island was diked and developed for rice cultivation. The impressive solitary brick stack, the single remaining construction from the Butler rice mill, gives some idea of the scale of the machinery needed to process the rice for market. It was Butler Island that English actress Frances Kemble made known to the world when her book, *Journal of a Residence on a Georgia Plantation in 1838–1839,* was published in New York and London in 1863.

The actress had captivated plantation owner Pierce Butler; and she married him in 1834 and began a quiet upper-class life in Philadelphia society. However, the press of plantation affairs, the source of Butler's wealth, and Fanny's desire to visit the rice fields eventually brought her to Georgia. By train, stage, and steamboat, the Butler party made its way down the seaboard to Charleston where they arrived on Christmas Day, 1838. After a brief stop in Savannah to purchase supplies, they reached Darien aboard the steamboat *Ocmulgee*. From Darien "two pretty boats" carried them across several "arms" of the Altamaha to Butler's Island.

The English-born wife of the plantation owner described her first approach to her exotic new home in the following passage:

> We now approached the low, reedy banks of Butler Island, and passed the rice mill and buildings surrounding it, all of which, it being Sunday, were closed. As we neared the bank, the steersman took up a huge conch, and in the barbaric fashion of early times in the Highlands, sounded out our approach. A pretty schooner, which carries the produce of the

estate to Charleston and Savannah, lay alongside the wharf, which began to be crowded with Negroes, jumping, dancing, shouting, laughing, and clapping their hands (a usual expression of delight with savages and children), and using the most extravagant and ludicrous gesticulations to express their ecstasy at our arrival.

Fanny seized the first opportunity to examine her new estate. Her colorful description of artisans at work forms a dramatic contrast with the "slimy, poisonous-looking swamp" from which the plantation had been wrested by dint of ingenuity and enormous amounts of hard labor:

> Pursuing my walk along the river's bank upon an artificial dike, sufficiently high and broad to protect the fields from inudation by the ordinary rising of the tide—for the whole island is below high-water mark—I passed the blacksmith's and cooper's shops. At the first all the common iron implements of husbandry household use for the estate are made, and at the latter all the rice barrels necessary for the crop, besides tubs and buckets, large and small, for the use of the people, and cedar tubs, of noble dimensions and exceedingly neat workmanship, for our own household purposes. The fragrance of these when they are first made, as well as their ample size, renders them preferable as dressing-room furniture, in my opinion, to all the china foot tubs that ever came out of Staffordshire. After this I got out of the vicinity of the settlement, and pursued my way along a narrow dike—the river on the one hand, and, on the other, a slimy, poisonous-looking swamp, all rattling with sedges of enormous height, in which one might lose one's way as effectually as in a forest of oaks.

Today, over 150 years after Fanny wrote them, her words evoke animated images of the slaves and planters to whom these low-lying islands in the Altamaha were once the vibrant center of a rice-dominated universe.

St. Simons Island

Unlike neighboring Jekyll Island, which was renamed when the English displaced the Spanish, St. Simons is the Anglicized form of the Spanish name for the island, San Simon.

Fort Frederica National Monument preserves and interprets the site of Oglethorpe's fortification and town begun in 1736 as part of England's power play to wrest control of the Southeast from Spain. The name honored the then-Prince of Wales.

Frederica's original settlers were a group of forty families brought from England by Oglethorpe. Almost as soon as they arrived, they set to work building an earth fort and laying out a town on a low bluff overlooking the river, a tidal creek on the west side of St. Simons Island. They also constructed two other outposts on St. Simon. With these blatant challenges to Spain's claim to hegemony in place, Oglethorpe was given British troops to man the forts; and he began preparing for war with the Spanish. The war came and is known in history as the War of Jenkins' Ear.

An incident in late 1739 gave Oglethorpe a reason for invading Florida; but the campaign was soon broken off by bad weather. The governor of Spanish Florida mounted a reprisal; it came in July 1742, in the form of a fleet of fifty-one ships and 3,000 fighting men. Oglethorpe withdrew his forces to Frederica and the Spaniards occupied the southern tip of St. Simons Island. On July 7th a force of invaders advancing toward Frederica were soundly beaten back by the British; and the same afternoon a British force of fifty successfully ambushed a large Spanish column at Bloody Marsh. The Spanish commander decided to return to St. Augustine with his main force intact. Never again did Spain mount a serious military challenge to England's presence on the coast north of Florida.

With the disbanding of the regiment in 1749, Frederica was soon nearly a ghost town and in 1758 a fire destroyed most of the remaining buildings. A stroll amidst Frederica's streets, building foundation excavations, and reconstructions, gives an idea of the character of the garrison town as it was 250 years ago. A short recorded narration at the site of the Battle of Bloody Marsh explains the historic significance of this live-oak-shaded site.

Close by the Fort Frederica entrance stands *Christ Church Frederica*. It is a reminder of the fact that the Anglican Church was the established church during the colonial period. It was at Frederica on March 9, 1738, that Charles Wesley entered the ministry. After he returned to England with his brother John, they played important roles in the evangelical revival that gave rise to the Methodist Church. The oldest gravestone thus far discovered in the churchyard bears the date 1803.

St. Simons Island, Georgia, to Savannah, Georgia

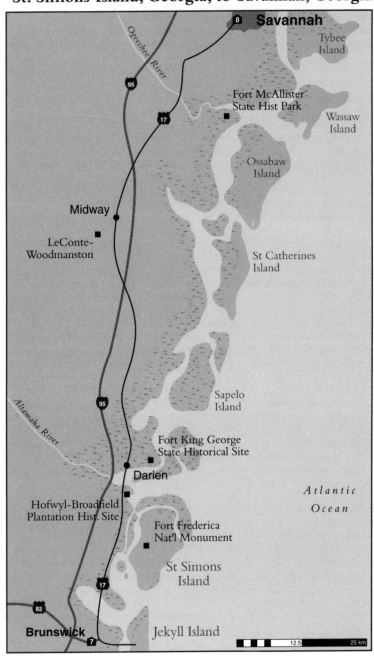

◬ Day Eight

ST. SIMONS ISLAND TO SAVANNAH, GEORGIA (85 miles)

Darien to Midway, Georgia, U.S. 17, 35 miles

Across the Altamaha is Darien, the historic county seat of McIntosh County, first settled by Scottish Highlanders. The town plan, an abbreviated copy of Savannah's, was laid out under the direction of Georgia's founder and architect, James Edward Oglethorpe. In a letter dated February 27, 1736, Oglethorpe wrote:

> I left St. Simon, rowing up the Alatamaha 3 hours I arrived at ye Scotch Settlement which they desire may be called Darien; They were all under Arms upon seeing a Boat, and made a most manly appearance with their Plads, broad Swords, Targets [shields] & Fire Arms. (Colonial Records of Georgia 21, 76)

It was not long before the Spanish began demanding that the English "should evacuate all they stood possest of as far as St. Helena Sound." Oglethorpe countered Spain's demand with one that challenged the limits of Florida. Then Oglethorpe rushed to Darien "and marked out the Fortifications for that place, and ordered a Church, School house and Guard house to be built." The solution to Darien's logistical isolation was provided when "Capt. Mackpherson arrived with a Drove of Cattel which he had brought all the way over Land from South Carolina." The herd's arrival "caused great Joy in all our Settlements, to find the Communication for Cattel by land opened whereby these Southern Settlements will be supplied with Milk and fresh Provisions of which they have hitherto stood in great need."

As the traveler enters Darien from the south over the US 17 bridge, the site of Oglethorpe's fortification is on the right. The historic marker indicates that it included "two bastions and two half bastions" in its layout. Immediately to the left, fronting on West Broad Street, is one of Darien's oldest buildings. Built about 1820, the tabby walls survived the fire and devastation that flattened much of the town in the Civil War. Other tabby ruins nearby date from the same period of cotton-export–related prosperity in Darien. As the traveler turns right on Ga 99, the Highlander Monument honoring the original Scottish settlers is on the left. Turn right on Franklin Street to *Vernon Square,* the original heart of Darien. The square is listed on the National Register of Historic Places. After touring the neighborhood around Vernon Square proceed east on Fort King George Drive for one mile to the fort.

Fort King George was built in 1721 by South Carolina Indian-trader Colonel John Barnwell. The visitor should view the twenty-minute slide presentation and exhibits in the museum prior to taking the self-guided tour of this historic site overlooking Black Island Creek and the Darien River branch of the Altamaha. The Fort King George site, with an authentically reconstructed blockhouse, is a memorial to the eighteenth-century imperial competition between France, England, and Spain for the American Southeast. In that competition the Indian Nations, as they were known, wielded political and military influence as they sided first with one and then another of the colonial powers whose ultimate objective was control of the land.

John Barnwell was sent to England to argue for the building of a series of frontier forts to forestall French expansion and serve as a network for British control over the Indian trade in the backcountry. As Barnwell pointed out to royal advisors, "the Method of the French is to build Forts on their Frontiers which it would be to our Interest to do likewise, not only to preserve our Trade with the Indians and their Dependence upon Us, but to preserve our Boundaries." Although they agreed with Carolina's grand scheme of imperial frontier outposts, the king's advisors funded only the post "at or near the mouth of the River Alatamaha in South Carolina."

Heading north on US 17, the tourist notices avenues of live oaks, hung with veils of Spanish moss, that line the road. A few words about their role in history are in order. Until fairly recently, ships were built of wood with relatively little in the way of metal fittings. Because of its resistance to dry rot, durability, and tensile strength, the semievergreen live oak (*Quercus virginiana*) that grows on the margins of the Coastal Plain from Virginia to Mexico found a valuable place in shipbuilding in the eighteenth and nineteenth centuries. It was the strength of the live oak used in building the U.S.S. *Constitution* that earned her the sobriquet "Old Ironsides." In the early history of the United States, Georgia provided the best live oak for the construction of the young republic's navy. Since there were no public lands on the Georgia coast, the navy contracted with property owners for the right to cut live oak. Fearing that foreign maritime nations would learn of the oaks' value and the tenuousness of its availability, Congress moved in 1799 to acquire forest lands in Georgia. Grover Island in nearby Camden County and the Blackbeard Island portion of Sapelo Island became the nation's first federally owned timber reserves.

Riceboro was an important town during the rice-producing era. It was first settled in the late eighteenth century and laid out in 1797. Riceboro functioned as the county seat of Liberty County until 1836 and was served by sailing sloops that could navigate the twenty miles of river channel from St. Catharines Sound.

Leconte-Woodmanston Plantation northwest of Riceboro is owned by the Garden Club of Georgia. The historic site is a portion of the original plantation acquired in 1760 by William and John Eaton LeConte, descendants of a French Huguenot who had migrated to America to escape the religious conflicts in his home country. John's son Louis assumed full control of the rice plantation in 1810 and devoted himself to creating a botanical garden that achieved international acclaim. Louis LeConte's sons, Joseph and John, were members of the science faculty of the University of Georgia in the antebellum period, then traveled west to help found the University of California. John served as that university's first president and Joseph was on the faculty and later helped found the Sierra Club. The visitor can pick out the location of the plantation's former reservoir and canal system used in the cultivation of inland swamp rice.

Midway

Some travelers may conclude that Midway was so named because it lies approximately halfway from Savannah to Darien on the old road. This is probably not the case since the nearby Midway River was originally named the "Midway," probably after the English river, before the settlement was established here. The settlement of the Midway district resulted from the migration in the 1750s of a large group of descendants of English Puritans. Their cultural and lineal ancestors had left Dorchester, England, in 1630 and founded Dorchester, Massachusetts. In 1695, a small congregation followed their minister to South Carolina and founded a Dorchester on the Ashley River near Charles Towne. When they moved to Georgia, the South Carolinians helped to expand the rice-culture frontier into that colony's tidal zone. In 1754 the newcomers formalized their presence by establishing a "Society Settled Upon Midway and Newport in Georgia." In a few years the transplants had brought in nearly a thousand slaves to clear and improve their Georgia land. By 1758, John Stevens, the largest planter in the congregation, had accumulated 2,000 acres that were being worked by thirty-five slaves. So successful were the original settlers at Midway that it was estimated that their plantations accounted for one third of the wealth in Georgia at the time of the Revolution.

While at Midway the visitor should examine the handsome *Midway Congregational* Church. It was built in 1792 to replace the 1756 building that had been burned by the British in the Revolutionary War. Particularly noteworthy are the original box pews. Next to the church is the small *Midway Museum* which will inform the traveler about some of the noteworthy families and individuals whose roots are here in Liberty County. Be careful in crossing the highway but don't leave without visiting the tree-shaded cemetery that was begun here in about 1754.

Midway to Sunbury, Ga 38, 7.5 miles

Leave Midway on Ga 38 heading east to Sunbury (round-trip about 15 miles). A slide presentation at the museum and visitors center located at the site of Revolutionary Fort Morris and War of 1812 Fort Defiance gives a valuable background for the visitor. Sunbury was named and laid out in 1758 by Mark Carr on a bluff site overlooking the calm waters of the Midway River. Carr recognized that the planters of the rapidly developing district would require a port for the shipment of their staple crops. In the early 1760s, the town began a period of rapid growth.

In 1773, William Bartram found Sunbury "a sea-port town, beautifully situated on the Main between Medway and Newport rivers." The nearly 100 houses in the town were "neatly built of wood having pleasant Piasas [sic] around them." According to Bartram "the inhabitants are genteel and wealthy, either Merchants or Planters from the Country who resort here in the Summer and Autumn, to partake of the Salubrious Sea Breeze, Bathing and sporting on the Sea Islands."

As might be expected, such an active port became an important target when war broke out with England. When they finally withdrew from Georgia, the English left almost nothing of Sunbury intact—the town had been destroyed and the social fabric of the district was in tatters. Some residents returned and rebuilt, but Sunbury never regained anything like its prewar stature. Its designation in 1783 as county seat insured some activity but even that was withdrawn in 1796 when the government was moved to Riceboro.

In its final years Sunbury functioned as primarily a summer resort for the families of the rice planters from the surrounding plantations. By 1841 the post office at Sunbury was closed, and in 1855 only a half dozen families lived in the old port. Today only an occasional shrimpboat or pleasure yacht is seen on the Midway where once a bustling seaport sent rice, cotton, naval stores, lumber, and leather to international markets.

Midway to Savannah, U.S. 117 or I-95, 35 miles

North of Midway on US 117, at the intersection of Ga 196, *Freedman's Grove* takes it name from having once been a village of freed slaves who received deeds to the land from their owner immediately after the Civil War. Although not common, there were several examples of such communities throughout the Plantation South.

At the intersection with Ga 144 is *Richmond Hill*. Here, in 1925, auto manufacturer Henry Ford began purchasing defunct plantations. By the 1940s he owned a total of 70,000 acres in Bryan and Chatham counties. Richmond Hill was developed as a model community based on the one at Dearborn, Michigan. But Ford came under fire from local historians when he had the Hermitage Plantation mansion taken down and the materials used to build a residence on his estate in Bryan County, and two of the Hermitage slave cottages taken apart and shipped to the Ford Museum in Dearborn.

Continue north to Savannah using US 117 or I-95.

△ Day Nine

SAVANNAH AND VICINITY

Shortly after he founded Georgia, James Edward Oglethorpe described the site of his capital, Savannah, in a letter addressed to the Trustees for Establishing the Colony of Georgia. The description was widely read since the trustees had it published in London's *Gentleman's Magazine* for April 1733. Oglethorpe wrote:

> I went myself to view the Savannah River. I fixed upon a healthy situation about ten miles from the sea. The river there forms a half moon, along the south side of which the banks are about 40 foot high and upon the top a flat which they call a bluff. The plain ground extends into the country five or six miles and long the riverside about a mile. Ships that draw twelve foot water can ride within ten yards of the bank. Upon the riverside in the center of this plain, I have laid out the town.... Over against it is an island of very rich land fit for pasturage, which I think should be kept for the Trustee's Cattle. The river is pretty wide, the water fresh. And from the quay of the town you see its whole course to the sea, with the Island of Tybee, which forms the mouth of the river; and the other way you may see the river for about six miles up into the country. The landscape is very agreeable, the stream being wide and bordered with high woods on both sides. (Mills Lane, ed., *General Oglethorpe's Georgia: Colonial Letters, 1733–1743*, pp. 4–5)

Savannah, Georgia and Surrounding Areas

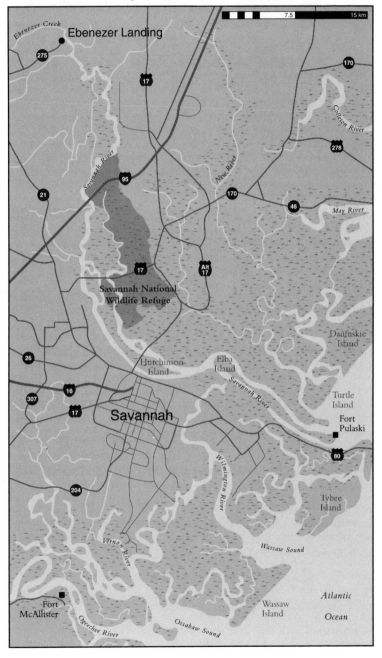

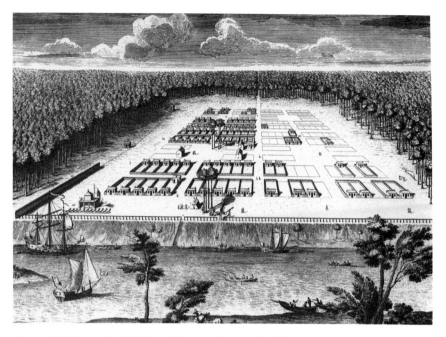

James Oglethorpe's "A View of Savannah . . . , 1734."

In the year that followed, Oglethorpe and his band of London artisans turned to the seemingly Herculean tasks of converting the pine barrens and swamps into a utopian colony. As a planned community took shape on the bluff overlooking the tidal waters of the Savannah River, Oglethorpe resolved to prepare and publish a view of the town. The product of Oglethorpe's resolve, titled "A View of Savannah as it stood the 29th of March, 1734," is considered to be one of the rarest of American urban prints. It presents an arresting high oblique perspective showing how Savannah's streets, unique squares, and building lots were cleared and surveyed in geometric regularity amidst the straight-trunked, long-leaf pines. The engraved "View" has assumed the status of icon in discussions of New World urban design, so widely published has it been in the present century. It showed only two-thirds or four of the six square-centered residential wards surveyed and cleared on the

Savannah, Georgia

sandy bluff the Indians called Yamacraw. Not until 1736 was a need felt to clear the lots in unshown Lower New Ward and Upper New Ward.

Several of the structures and scenes portrayed in "A View of Savannah" invite discussion. For one, the steep escarpment of the bluff requires stairs and a handrail for passage up and down. Immediately to the right of the stairs can be seen the "crane" that the colonists constructed to hoist heavy barrels, boxes, and bales up the bluff escarpment with the aid of a steeply inclined ramp of timber skid rails.

Prominent among the constructions shown is "The Pallisadoes." This depiction of an incomplete log fortification has led some experts to conclude that Savannah was originally built within a log palisade. Such was not the case. The palisade was begun "round the place of our intended settlement as a fence in case we should be attacked by the Indians." A short segment is all that Oglethorpe ordered built because he soon discovered that the only Indians within fifty miles were not only "at amity but desirous to be Subjects to King George, to have lands given them amongst us and to breed their children at our Schools."

Although Oglethorpe and his colonists were unaware when they laid the city out, Savannah was destined to be located squarely in the tidal rice zone along the river from which it took its name. By the time of the Revolutionary War, large areas of riverine swampland and marsh islands in the vicinity of Savannah had been improved for rice cultivation.

The expansion of rice culture in Georgia gave rise to most of Savannah's foreign trade that grew so steadily after 1760. In 1772, Savannah's rice export totalled 23,540 barrels, and was worth £69,266. Damage and neglect took its toll on the rice plantations during the war, but the greatest impact on Savannah's economy came from several sunken wrecks which blocked navigation lanes and threatened to destroy the port. As these problems were addressed there was no lessening of interest in tidal rice production along the Savannah River.

But as the swamps and marshes were converted into productive flooded rice fields, some inhabitants began to voice concerns.

Mists called "vapors" and "miasmatas" could often be seen forming above the fields and wafting toward the city. It was felt that the tall trees, now cleared away, had formerly caused these deleterious vapors to rise and dissipate in a harmless fashion. William Gerard DeBrahm, longtime Savannah resident and surveyor general of lands, described this in his 1772 Report:

> The City of Savannah . . . for near 30 years was accounted a very healthy Place. The South Carolinians used to come there for recruiting their Health, but the Vapours which generally breed in Swamplands and rise by help of its high Trees, as through Chimneys, to gain the free Air, in which by any moving Wind they are carried the Road of the Wind. . . . but as soon as the Trees on both these Lands were cut down and the cleared Land converted into Rice Fields, the Vapours hanging upon them at present are by a North or East Wind (for want of their living Ladders to ascend by them above the Summit of the City) rolled in it, and all the Streets and Houses filled with them, to the Prejudice of its Inhabitants, whose Diseases are in every respect similar to those in the Neighbouring Province of South Carolina. . . .

Perceptions such as DeBrahm's were by no means unusual in eras preceding the general adoption of the germ theory of disease cause and spread.

What is truly extraordinary is the way the people of Savannah responded to the perceived threat. By the late eighteenth century Savannah's physicians and civic leaders were generally agreed in the belief that miasmata originating in fetid rice fields and other poorly drained land were the major cause of the often-fatal diseases known as "autumnal fevers." In 1804 the city's physicians came together to found Georgia's first medical society to help lessen "the fatality induced by climate and incidental causes" as well as to improve the science of medicine generally. This group adopted the idea that Savannah's public health could be greatly improved by draining the offending fields near the city. On March 24, 1817, Savannah enacted an ordinance providing for the first long-term program in the United States for the control of disease through a large-scale and costly modification of the city's surrounding environment through land drainage.

It was recognized that much of that land could only be kept dry by ending a lucrative system of rice cultivation and thus by inflicting severe economic penalties on affected plantation owners. Contracts were drawn up requiring landowners to undertake "the alteration of the culture of said land from a wet and water-flowing culture to a dry one as being more salubrious and beneficial to the health of said city." A maximum compensation of forty dollars per acre was paid for the losses to the landowners under the scheme.

Approximately three square miles of land adjacent to Savannah were converted from wet to dry culture as a result of the 1817 enactment. A sum of $70,000 was appropriated in 1817 to compensate the affected landowners, with an additional $15,000 being required in 1819. The residents of Savannah paid an average cost of about fourteen dollars a person to put some 10 percent of the Savannah River's tidal rice land out of production. Through the period of the 1820s, this experiment was perceived as successful in lowering Savannah's autumnal-disease death rate.

One of the unforeseen results of Savannah's bold attempt at environmental regulation for public health benefit was to hasten the city's ability to embrace the large-scale transportation and industrial activities in final decades of the antebellum period. Landowners restricted to dry culture and in control of acreages immediately adjacent to the city and most in demand were eager to sell their dry land for development.

One of the best places to begin a visit to Savannah is the *Savannah History Museum* near the Savannah visitors center at the western edge of the historic district. The museum charges a small fee for admission. There a colorful film presents Savannah's history from its founding to the present day. For military-history buffs, there is an audio-visual presentation of the 1779 Siege of Savannah, probably the most decisive victory Britain enjoyed in the Revolutionary War. The museum building itself is a historic structure that was once the center of Central of Georgia Railway operations. It represents a prime example of the adaptive use of historic buildings, a development Savannah helped pioneer. On leaving the museum, the visitor should walk a short distance south along Martin Luther King Boulevard to the Georgia Roundhouse Complex where the fascinating history of Georgia's railroads is displayed and interpreted.

After a driving tour of the streets and squares of Savannah's world-famous historic district, travelers can complete their morning by visiting one or more of the several houses that are open to the public. Finding an interesting venue for lunch is similarly a matter of individual preference because Savannah offers so many choices to fit every appetite and pocketbook. Brochures available at the visitors center will be helpful in assuring that memorable choices are made.

Ebenezer

To reach the German ghost town New Ebenezer on the Savannah River, leave Savannah by Ga 21 heading toward Springfield. Turn right on Ga 275 and continue to the end of the road. Ebenezer is an example of a colonial settlement that became a viable and successful producer of agricultural produce, including silk and rice, without adopting the slave-based plantation system. How such a community managed to come into being is a story worth telling in the shadow of historic Jerusalem Lutheran Church, the one remaining structure in Ebenezer and Georgia's oldest still-functioning building. The church was built in 1767 with handmade bricks, some of which still show impressions of their maker's fingers.

At about the time that plans were being made to found the new colony of Georgia, the Catholic archbishop of Salzburg issued the infamous Edict of Expulsion. By its terms, all Protestants were forced to leave their Salzburg homeland. About 20,000 Evangelical Lutherans found their way to East Prussia.

King George II of England, who was also the Duke of Hanover-Prussia, was deeply moved by the plight of the Salzburger exiles (in his public role the king served as official head of the Anglican church, but in private life he remained a staunch Lutheran). Not surprisingly, groups close to the Crown such as the Trustees for

Establishing a Colony in Georgia actively espoused the Salzburger cause. Oglethorpe and his fellow trustees viewed hard-working German Protestants as highly desirable potential settlers for their planned buffer-zone colony. To this end, in 1732, they commissioned Samuel Urlsperger, senior of the Lutheran ministry in Augsburg, to recruit 300 Salzburgers for settlement in Georgia.

The first "transport" of Salzburgers for Georgia fell far short of that number. About twenty-five, already en route to Prussia, were persuaded to change their destination to the newly established colony. Several more joined the group which was under the charge of a young Hanoverian nobleman, Philip Georg Friedrich von Reck. Upon their arrival at Rotterdam the transport was joined by the two young Lutheran pastors who would control almost every aspect of the settlers' lives in their new homeland.

On March 5, 1734, after a cramped and difficult eight-week passage aboard the *Purrysburg,* the Salzburgers made a landfall on the Carolina coast. They were given a festive welcome in Savannah on March 12, 1734. Part of the group set out to select a site for the Salzburgers to settle. Ostensibly the Germans were free to choose their own location. In fact, Oglethorpe probably influenced them to choose a site that would strengthen the defensive perimeter around Savannah.

In view of the sandy, virtually rock-free character of the Salzburger's chosen site, it is somewhat ironic to find that they named their settlement Ebenezer, meaning "Rock of Help." Religious zeal rather than careful terrain and biota appraisal was obviously the chief force motivating Georgia's original German-speaking settlers.

The rigors of pioneer existence and the scourge of dysentery began to take their toll. Struggle as they would, the Salzburgers were no fit match for the bareness and harshness of the site they had settled on Ebenezer Creek. What had first been thought to be fertile soil was discovered to be a thin veneer of black humus masking white sand.

While foraging for food for their hogs, the Salzburgers encountered a stand of oaks on a clay bluff overlooking the Savannah River. Von Reck wrote in his *Journal* of how he and the pastor:

informed Mr. Oglethorpe of the conditions of the Salzburgers, the nature of the soil at Old Ebenezer and the difficult communications between this place and the other English plantations and had suggested for a real and profitable development of the city of Ebenezer a good and fertile place at the confluence of the Ebenezer and Savannah rivers.

Oglethorpe agreed to moving the Salzburgers to the Savannah River at Red Bluff. After the site was surveyed and divided, a lottery was held to assign house lots and the Salzburgers set to work building houses of "a simple yet comfortable style of architecture."

Ebenezer was hard hit by the military action of the American Revolution. It was occupied by British regulars under the command of Col. Archibald Campbell in 1779. In a report sent to Lord George Germaine on January 20, 1779, Campbell mentioned "some redoubts" he was having constructed at the important Ebenezer crossing point. He explained that he intended to make Ebenezer "the Advance Magazine for the Army." On March 2, 1779, he admonished General Prevost that, "the security of Savannah, Ebenezer and Sundbury . . . ought never to be neglected" whatever enemy movements might be. In the same communication Campbell mentioned that 1,390 men would be defending Ebenezer.

In 1782, Brigadier General Anthony Wayne led an offensive in Georgia and drove the British out of Ebenezer and other outlying settlements. In July of that year Georgia's legislature met in Jerusalem Lutheran Church at Ebenezer, in a technical sense making the town the state's temporary capital. About the same time the British forces finally evacuated Savannah.

Ebenezer never recovered as a town after the British occupation. Many of the houses and buildings were damaged or destroyed during the war years. Rather than rebuilding on town lots many of the settlers chose to build on outlying farmlands. After visiting Jerusalem Church and the small Salzburger museum nearby, the traveler should stroll through the pines to the bank of the Savannah River. The earth revetments of the Revolutionary War fortifications can be traced by walking through the woods on the landward side of the town site. After visiting Ebenezer, the traveler should return to Savannah.

The Historic Savannah Waterfront

The present-day merchants and restauranteurs who comprise the Savannah Waterfront Association correctly boast that their colorful riverfront precinct is "Where All Georgia Began." As we walk on cobblestones within sight, sound, and smell of the river, we are at the foot of the sandy escarpment where the sailing craft drawing "twelve foot water" shown in the 1734 engraved "View of Savannah" could "ride within ten yards of the bank."

In this part of its course the water of the Savannah is split by Hutchinson Island, the shoreline immediately to the north. The water between Savannah's waterfront and the island is called Front River, and the channel between Hutchinson Island and the South Carolina shore is called Back River. As might be expected the Georgians devoted themselves to maintaining the navigability of Front River.

As Georgia became a royal colony and plantation agriculture began to flourish, complaints concerning the inadequacies of Savannah as a port were more frequent. In 1773, Governor Sir James Wright confirmed earlier complaints by pointing out the hazards posed by "three sand banks in different Places" in the river downstream from Savannah. Until they are removed, the governor added, "Vessels at the Town do not load deeper than from twelve to thirteen feet and then are obliged to fall down to Cockspur to take in the rest of their loading." Cockspur was an island near the mouth of the Savannah River and thus exposed to wind and waves.

In spite of these drawbacks, Savannah flourished as a port and attempts to shift Georgia's capital to the Ogeechee River failed. Improvement of the Savannah River's navigability became a major preoccupation of city, state, and finally federal authorities. These efforts continue today in a variety of channel improvement schemes and, most visibly, the new high-level highway bridge spanning Front River just upstream from the historic waterfront.

In 1844 a Scottish Presbyterian minister named George Lewis toured America and published his observations in a book titled *Impressions of America.* Lewis was very favorably impressed by Georgia's chief port and largest city. "Savannah is the greatest export town of the South, greater than Charleston, and not much inferior to New York," he averred in a burst of enthusiasm that must have endeared him to Savannahians reading his book.

Perhaps the traveler should now compare her or his own impressions of Savannah's sights and delights with those of Swedish scholar C. D. Arfwedson, who came to the South in 1833 and wrote:

> The first view of Savannah produces no very favourable impression . . . ; but upon nearer inspection, he cannot avoid pronouncing a fair opinion of the city, namely that it is far from unpleasant, and that the houses of the inhabitants are ever open to strangers, to whom all the attentions of hospitality and politeness are shown. Savannah is situated on a ridge of sand, close to a river bearing the same name, which divides the States of South Carolina and Georgia. . . . In that part of the walk contiguous to the river, as well as in the one below on the bank itself, all business is transacted, which consists chiefly in cotton shipments to Europe. In this particular spot, nothing is heard but conversations about the article; bales are piled up in every store and at every corner. Whoever visits a merchant's office will find it filled with samples: if the clerks are occupied with correspondence, rest assured the subject is cotton. If the chief of a mercantile house is seen in conversation with any one, be equally sure that he is talking about the price of cotton. But in the interior of the town reigns perfect tranquillity; nothing indicates that on the article of cotton alone depends the prosperity of Savannah.
>
> The houses are mostly of wood, and have balconies in the usual southern style. The unpaved streets resemble well-kept high roads. After heavy rains, or, more properly, drenching showers, it is customary to plough the ground, as in a field, with a view to render it sooner dry. This method of drying streets appeared rather new to a stranger; but certain it is that the object was accomplished with extraordinary rapidity, for the depth of the furrows enabled the sandy soil to imbibe the water much sooner than if it had been left to drain away of itself.

Generalized Areas of Indigo Production, South Carolina

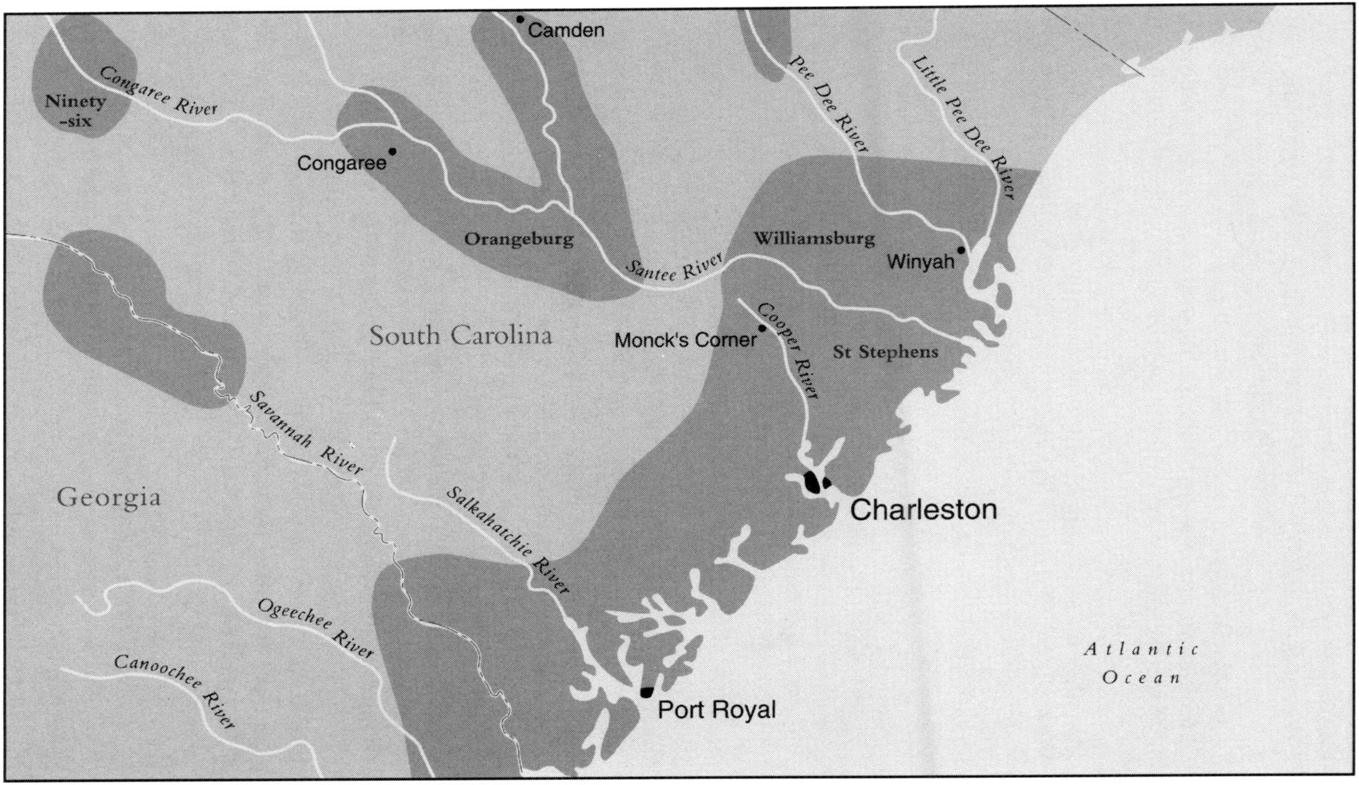

Since Arfwedson's day many things have changed: cotton candy has replaced cotton bales and the sandy streets are paved; but historic Savannah remains a city with perfectly straight streets and squares where "all the attentions of hospitality and politeness are shown" to the visitor.

Indigo, Pioneer Staple Crop on the Rice Coast

Some of the original settlers in South Carolina included indigo among the cultigens they brought with them from their West Indian island homes in 1670. Although their experiments with the dye-producing plant were moderately succesful, indigo failed to become a staple crop. Soil exhaustion, droughts, severe winters, inadequate manufacturing expertise, poor marketing, and low prices were among the factors contributing to the initial commercial failure of the South Carolina indigo industry.

A second and more successful era of indigo production owed much to the persistence of the daughter of a South Carolina plantation owner, Eliza Lucas. Writing to her father in 1743, after descriptions of crop failure she added, "I have no doubt Indigo will prove a very valuable commodity in time." The following year, Eliza's persistence paid off in the form of a successful crop.

During the period of her experiments and travails, indigo was being experimented with in neighboring Georgia and North Carolina. When England's normal sources of supply in the French and Spanish West Indies were cut off by the conflict known as King George's War (1739–1748), the crop was moderately successful. The same period of conflict saw maritime insurance rates climb to a point that discouraged the shipping of bulky commodities like rice, and Rice Coast planters turned much of their effort toward the cultivation of indigo. In 1747, 138,300 pounds of the blue dye stuff was exported from South Carolina.

The return of peace brought a recovery in rice profitability but interest in indigo was maintained by the news that Parliament had provided a bounty of sixpence for every pound of American indigo shipped into England, effective March 28, 1749. The French and

Indian War (1754–1763) brought an unprecedented boom in indigo production. South Carolina geographer John Winberry describes the distribution of the crop in his article "Indigo in South Carolina." (*Southeastern Geographer* 19, no. 2, Nov. 1979, 91–102). The article reflects the spread of indigo cultivation into the Piedmont. Although soil-demanding and labor-demanding in its growth and preparation, indigo itself was a relatively high-value–low-bulk commodity that could bear the transport costs of wagon haulage to coastal ports.

Savannah, Georgia, to Charleston, South Carolina

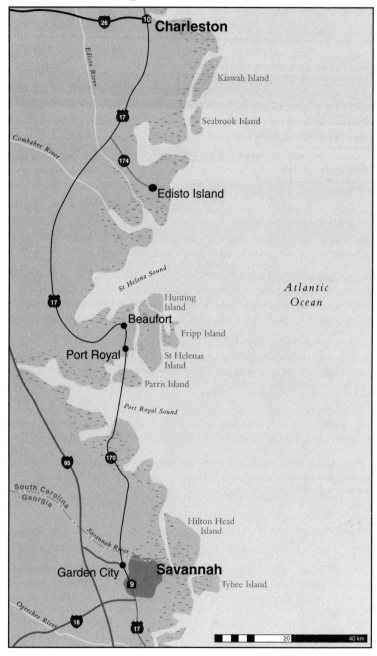

Day Ten

SAVANNAH, GEORGIA, TO CHARLESTON, SOUTH CAROLINA (140 miles)

Savannah, Georgia, to Hilton Head Island, South Carolina, 40 miles

Take US 17 north from Savannah. The road goes through *Yamacraw,* site of the Indian village of Oglethorpe's Indian friend Tomo-Chi-Chi, with whom the arrangement was made for the first English settlement at Savannah. This area is a residential area for African Americans; and the west side of Bay Street Extension shows the low cost housing, built during the 1930s, which replaced the dilapidated shacks that formerly characterized the area.

The *First Bryan Baptist Church,* easily seen from Bay Street, is the oldest African-American church in the United States. It was established in 1788 by Andrew Bryan, a slave of Jonathan Bryan of Brampton Hall. According to tradition, Bryan was whipped by Savannah magistrates for preaching to blacks in Yamacraw without white supervision; but later the city granted him a lot for a meetinghouse. The church moved to the present site in 1793 and the existing edifice was built about 1873. Sunday services are held at 11:00 a.m., and, according to the church announcement, "visitors and tourists are welcome."

Before the Civil War, blacks went to white churches, but with segregated seating. They nevertheless showed a preference for their own preachers and worship style, and after the war, African Americans quickly moved to establish their own separate churches.

About 3 miles north of town is the Union Bag and Paper plant—one of the first plants to be erected in the South to apply the results of Dr. Charles Herty's experiments in the manufacture of paper from Georgia slash-pine pulp. In the Southern pine economy, the manufacture of brown bags and cardboard has replaced turpentine and other naval stores as a major tree crop.

The plant is located on the site of Hermitage plantation. In the early nineteenth century, this was the site of Henry McAlpin's brick kilns. The dull red brick, known as Savannah Grey, extensively used in brick construction in Savannah, was made here.

About four miles farther north a spur highway leads to the Savannah State Docks of the Georgia Ports Authority. Here, east of Garden City, is the large container facility, developed since 1945, which has helped to make Savannah the leading Southeastern port.

Next the highway passes through Port Wentworth, originally Joseph's Town, one of Savannah's original outlying settlements. About 3 miles north of here was the Mulberry Plantation, granted by the State of Georgia to Revolutionary hero General Nathaniel Greene. Here young Eli Whitney, tutor to the Widow Greene's children, invented his new type of cotton gin.

Across Front River, the main channel of the Savannah River, is Onslow Island, the beginning of the 27,000-acre *Savannah National Wildlife Refuge*. It is one of the seven refuges on the Georgia-South Carolina Coast spanning a distance of a hundred miles and including over 53,000 acres with both saltwater and freshwater habitats. Pollution control has become a major problem since the refuge was established in 1927. Thirty-two major industries are located on Front River just north of Savannah.

Continue north across *Argyle Island,* made famous by the plantation accounts of the Manigault Family of Charleston, South Carolina. Argyle Island is separated from the mainland by Little Back River, the third channel of the Savannah at this area. It forms the boundary with South Carolina. Leave Chatham County, Georgia, and enter Jasper County, South Carolina. About one mile from Back River is the headquarters of the Savannah National Wildlife

Refuge. The headquarters offers no tourist services except for a display rack of brochures. The refuge is administered by the U.S. Fish and Wildlife Service of the Department of the Interior.

Thirteen rice plantations once covered the tidal freshwater area near the refuge headquarters. Opposite the headquarters building is the entrance to Laurel Hill Wildlife Drive, a four-mile drive along retaining dikes which gives a good view of the holding ponds and the diversity of wildlife. The most common birds seen throughout the year are long-legged waders such as egrets, heron, and ibis.

Laurel Hill takes its name from a 400-acre rice plantation purchased in 1813 by James Hansell Acrum. The live oaks on the border of the refuge knoll once fronted the plantation house. The complex included slave cabins for 150 slaves and a steam-powered rice mill. The brick wall and floor of the mill are the only structural remains of the plantation; and the millstone at the entrance to the small cemetery is thought to have been used here.

Return to US 17 and continue north to SC 140; turn right again on US 46 to Bluffton on the May River.

Bluffton began as a summer resort for inland rice planters before the Civil War, when the barrier islands were still inaccessible. Still popular as a saltwater retreat, the town has a small antebellum Episcopal church in Gothic Revival style.

To reach *Hilton Head,* a barrier island overrun with resort development, continue on SC 46 from Bluffton to US 278 and turn right; the island lies six miles ahead. Hilton Head, with an area of forty-two square miles, is one of the largest of the barrier islands. It takes its name from Captain William Hilton, the captain who explored the Carolina coast in 1663 on behalf of Barbadian planters contemplating a move to Carolina. His reports helped to shift interest from Cape Fear to Port Royal Sound.

Since 1956, when bridges were built across the Intracoastal Waterway, the island has been accessible by highway. Prior to that time, the island was inhabited by only a few fisherman and black farmers, eking out a living. Today it is a highly developed seaside resort, with a large resident population as well as a million transients a year who come for golf, sun, and surf. The original residents of the area have not profited from development as much as outsiders, a reminder that when local owners lose their land to "progress," they may also give up their own way of life. Resort development has drastically altered both the economy and landscape of this part of the low country.

Hilton Head to Parris Island, 34 miles

Return from Hilton Head to SC 170 and continue north. The highway crosses the Broad River at the head of Port Royal Sound. Turn right onto SC 802 and go to the intersection with SC 280. Turn right and continue to the main gate of the U.S. marine base at Parris Island. At the entrance gate, obtain a visitor's permit from the sentry, then proceed past the Iwo Jima Monument and the parade ground to the visitors center to pick up a driving tour brochure and map of the depot. Go directly to the museum to see artifacts from the Santa Elena excavations, and then follow the road over Ballast Creek to the historic sites.

Santa Elena was established in 1566 to be the capital of a North American colony. After Indian attacks in 1576, it was evacuated; but the Spanish returned and built a larger settlement and Fort San Marcos. This lasted until 1587, when British raids forced a contraction in Spanish settlement. A Spanish mission, however, was maintained in the Port Royal area until the middle of the seventeenth century, as one of the Guale (Georgia) missions.

Cross bridge over Ballast Creek and pass Page Field; at the first intersection, which leads to historic sites, is a parking circle.

Concrete blocks mark the outline of Fort San Marcos. In the center is the *Jean Ribault monument*. When it was erected in 1925, this was the presumed site of Charlesforte. More recent archaeological excavation has definitely established that this was a Spanish, not French, site. The exact location of 1562 Charlesforte has never been ascertained, although a probable site is on Ballast Creek.

North over the golf course is the site of Santa Elena I (1566–1576) and Santa Elena II (1577–1587). Between 1577 and 1587, the view of town would have been obscured by the high parapet of earth from the wide moat that encircled the fort on the landward side. Despite 350 years of erosion, when the earthworks were razed in 1917 they were still ten to twelve feet high, and the outlines of a wide moat were still traceable. About 100 yards to the east is the site of San Felipe II, the fort which protected Santa Elena I. The two southern bastions have washed away, suggesting how erosion could have altered the typography of Parris Island.

The tip of Parris Island is a good place to view *Port Royal Sound*. To the east a creek made Santa Elena accessible by small craft from the Beaufort River. To the west is the Broad River; to the south is Port Royal Sound and the entrance to the Atlantic. Flanking the entrance are two islands on which Confederates had erected earthen forts: Fort Beauregard on Phillips Island to the east, and Fort Walker on Hilton Head to the west. (After Union capture, changed respectively to Fort Seward, for the secretary of state, and Fort Welles, for the secretary of the navy.)

On November 4, 1861, Rear Admiral Samuel F. DuPont arrived off Port Royal bar with a fleet of forty-five vessels. Two days were spent in laying marking buoys, scaring off Confederate Commodore Tattnall's collection of vulnerable riverboats, and perfecting attack plans. The engagement began on the morning of November 7 (the wind on the 6th had been too boisterous for maneuvers) and shortly after noon Confederate forces began to withdraw from Fort Walker. A boat sent ashore under a flag of truce at two o'clock found that Fort Walker had been abandoned. The Confederates also evacuated Fort Beauregard, leaving their armament behind. Later it was ascertained that the Confederates had withdrawn from all their coastal defenses between Edisto Island and the Savannah. The way was open for Port Royal to become the headquarters of the South Atlantic blockading fleet, in accordance with Union plans.

Visitors may return to the main gate by Wake Boulevard or visit other sites marked on the visitors map, such as Elliots Beach and the historic district, including the commandant's Victorian house and the remains of the drydock, on which construction began in 1891. Parris Island takes its name from Colonel Alexander Parris, public treasurer of South Carolina before 1733.

THE PLANTATION SOUTH

Parris Island to Beaufort, 10 miles

Leaving the marine base by the main gate, take SC 281 into Port Royal, at the tip of the point between Battery Creek and the Beaufort River. The South Carolina Ports Authority maintains a small port facility here.

Historically, Port Royal lacked river access to the interior. The two main interior rivers nearby—the Combahee and the Coosawhatchie—were both low-depth swamp rivers navigable in the days of riverboats primarily in the thinly populated piney woods area of the Coastal Plain. In the 1870s and 1880s, there was an increase in Port Royal activity associated with phosphate mining and exportation, but this activity was not revived after the killer hurricane of 1893.

The Penn Center on Land's End Road, St. Helena Island, is a legacy from the period of the Port Royal Experiment, a prelude to Reconstruction, which included charitable efforts to educate the freedmen, help former slaves make the transition to a wage economy and prepare them for citizenship, and sell acreages of confiscated land to blacks. The dreams of the freed floundered when their outside benefactors became more interested in the imposition of their own ideas of social reform on the former slaves than in helping them achieve true independence.

That the Penn effort survived was due largely to the work of two Pennsylvania women, Laura M. Towne and Ellen Murray. Towne, a fervent abolitionist from Philadelphia, answered a call for teachers and was among the dedicated group that arrived by steamer on April 17, 1862. She was later joined by her friend Ellen. They named the prefabricated school building sent to them the Penn School because of the great support given to black education and relief by charitable organizations in Pennsylvania.

After the war was over, interest in aiding the blacks declined, and only with great difficulty was Towne able to keep the Penn School going. But she had the satisfaction of seeing two generations of black children educated there. When state schools were finally established in the 1870s, they were taught by black teachers, most of whom had received training in the Penn School. Over

ST. HELENA ISLAND
(Round trip about 50 miles)

Return to SC 281 then turn east at the intersection onto SC 802; cross the Beaufort River to Ladies' Island. At the intersection with US 21, turn right to Frogmore on St. Helena Island. Continue west to the Penn Center, an African-American cultural-heritage center.

St. Helena Island is the largest of Beaufort County's sixty-five islands. Described geologically, St. Helena is an erosion-remant island, formed when tributaries of large streams cut new paths to the ocean. While many waterways along the ocean in Georgia and South Carolina are called rivers, they are in reality ocean arms, bringing salt water close to the pine belt, often without meeting any significant volume of fresh water.

There are two other types of coastal islands. Marsh islands are formed from the accretion of tidal sediment and wind-driven sand, especially on the ocean side of former mainland. Beach-ridge, or barrier islands front the ocean, where the endless dynamism of wind and wave constantly carry away or bring in new sand, tearing down and building up. Four barrier islands stand between the ocean and St. Helena.

For many years after the Civil War, St. Helena and other low-country islands had a majority black population. They lived in relative isolation from whites and retained many distinctive beliefs and practices. With an economy based on getting a livelihood from the sea and small-scale farming, they gave to the low country a distinctive ambience.

Since World War II, great growth in tourism and in retirement communities brought an increase in the white population and a decrease in the black population. Today the major agricultural activity on St. Helena Island is tomato farming, but many locals make early morning bus runs to work on booming Hilton Head Island, just across Port Royal Sound but a hundred-mile round-trip by road.

Prior to World War II, a coastal dialect known as "Gullah" survived from Georgetown, South Carolina, down through the Georgia coast, where a less extreme variant known as "Geechee" was the local speech. There was always a great variation in the patois, depending on the degree of isolation. Today little Gullah survives, and outsiders are unlikely to encounter it spoken except in cultural-heritage centers.

Linguistic studies now indicate that the origins of Gullah lie in the emergence of a pidgin in Africa for the purpose of general trade and slavery. It was intensified by the practice of distributing new slaves among already seasoned hands. In order to communicate, they were forced to learn the pidgin in use; whatever English was learned by field hands was by the process of slave transmission. With increasing black-white separation, a distinctive form of black English developed, which affected the white speech of educated whites, in intonation and stress, if not in syntax.

---

time the curriculum of the Penn School was enlarged to include industrial and agricultural training.

Long before the Civil Rights movement, Penn Center was a center for African-American inspiration, and one of its visitors in the 1960s was Martin Luther King, Jr.

Today the Penn Center presents a decaying cluster of buildings under the spread of great live oaks, which gave the name The Oaks to this antebellum plantation. The Penn Center is listed by The National Trust for Historic Preservation as an endangered historic site; but demographic and economic changes have made Penn Center less important in African-American development. Today the center still promotes community services. Its York W. Bailey Museum focuses on the Sea Island black heritage. A donation is requested.

Before proceeding on to Beaufort, visitors may wish to continue west to Port Royal Sound. East on Seaside Road lies *Tombee,* the oldest standing house on St. Helena. Dating from 1795, this private residence is a two-story house on high pillars, with a double piazza across the front. It has been well restored. Tombee became famous as a Sea Island cotton plantation in 1986 with the publication of Theodore Rosengarten's *Tombee: Portrait of a Cotton Planter,* which contains the plantation journal of its owner Thomas B. Chaplin (1822–1890).

Return to US 21 and turn north to Beaufort via Ladies Island. This island was claimed by the Spanish on Our Lady's Day in 1525, and their name for it was retained by the British.

Beaufort to Charleston, 56 miles

Located on Port Royal Island on a bluff overlooking the Beaufort River, Beaufort is reached from the south from Ladies Island by the Beaufort River Bridge, completed in 1927. The Beaufort visitors center, at the foot of Bay and Charles streets, is a convenient place to pick up maps and plan a drive around Beaufort. Much of the charm of the city lies in its great trees and gardens with vistas over the Beaufort River.

The lords proprietor gave Lord Cardross a grant to make a settlement of Scot Covenanters here in 1684, at a site named Stuart Town. The settlers unwisely incited Carolina Indians to attack the Guale Spanish missions, and in retaliation the Spanish destroyed the embryonic town. Not until 1710 was the present Beaufort (locally pronounced Bew' fut) laid out and settled, this time by settlers with previous experience in Barbados and other colonies.

The new settlement was named after Henry, Duke of Beaufort—one of the proprietors. It got off to a good start, but was almost wiped out in the Yemassee War of 1715. This war began at Pocotaligo, a nearby Yemassee Indian town, whose residents felt threatened by new settlement and angry over trader abuses. In the Revolutionary War, Beaufort was captured by the British who burned the properties of the most noted Revolutionists of the area.

THE PLANTATION SOUTH

In the colonial period, indigo was Beaufort's major export staple; but Sea Island cotton replaced it in the nineteenth century.

More of a resort town than a city, Beaufort became known for summer residences for local cotton and rice planters, following a pattern set by Charleston. The planters' town houses were often more elaborate than those on their working plantations. Beaufort architectural styles are mixed, from Charleston-style to plantation-plain and Federal.

On the abandonment by the Confederates of the forts on Port Royal Sound on November 7, 1861, the whites abandoned Beaufort and the adjacent islands, leaving them to the Union forces and slaves. The following account was written by Daniel Ammen, Rear Admiral, U.S.N., in 1887 for *Battles and Leaders of the Civil War:*

> On the following day [November 9, 1861] . . . [the *Seneca*] was sent to Beaufort, . . . On the wharves were hundreds of negroes, wild with excitement, engaged in carrying movables of every character, and packing them in scows. As the gun-boats appeared, a few mounted white men rode away rapidly. A very beautiful rural town had been abandoned by all of the white inhabitants, quite as though fire and sword awaited them had they remained. Instead of that, I was directed by the flag-officer to assure the peaceable inhabitants that they would be protected in life and property. This message was delivered to the only white man found, who sat in the post-office and seemed quite dazed. At General Drayton's headquarters [commander of Confederate land forces] was found a chart of the coast and, in red pencil marks, a very valuable addition, no less than the position of all the earth works in his command, the number of guns being shown by the red marks in each locality. All of the batteries from North Edisto south to Tybee were found to be abandoned; the guns, however, had been removed, with the exception of some inferior pieces. Wherever the gun-boats penetrated, into harbors or rivers, huge columns of white smoke were seen on all sides from the burning cotton, far out of our reach, had it not

been the special object of our visit to secure it. Thus the enemy inflicted upon the inhabitants injuries they would have otherwise escaped, even had it been within the power of the crews of the gun-boats to inflict them. (I, 688-689)

Abandonments of plantation and town property meant that the owners were unable to pay the Federal taxes imposed, and most of the property was confiscated and sold to Northern speculators, who poured in on the heels of the army; some lands were reserved for sale in small lots, undersized for profitable farming. The hurricane of 1893 severely damaged the area (as did one in 1940), and during the early twentieth century, agriculture declined sharply. More and more land passed from cultivation into hunting preserves for wealthy Northern sportsmen. The new economy is based on tourism.

Beaufort's 300 acres of old homes and buildings was declared a National Historic Landmark in 1974 and is remarkably free of shopping centers and fast food restaurants, which form an almost endless strip on approaching highways. Except for two former residences open to the public as house museums, the homes of Beaufort are privately owned. The town is a place to be savored by walking or bicycling. On foot a visitor can take in the total ambience, appreciate architectural details, and enjoy the storybook setting of trees, gardens, and mansions. Visitors interested in crafts and antiques will find a number of good shops in the two-block Bay Street.

On the way out of town US 21 passes the *National Cemetery,* established 1862. It contains the graves of 12,000 Union soldiers who died in the South. Continue north to Gardens Corner where US 21 meets US 17.

Gardens Corner is named for Colonel Benjamin Garden, a former landowner; although the name has been attributed to Dr. Alexander Garden, a prominent physician and the naturalist for whom the sweet-scented gardenia is named.

At Gardens Corner turn right on US 17 toward Charleston. The road crosses the Combahee River, in the area where the Heyward family once had extensive rice plantations. Duncan Clinch Heyward, governor of South Carolina 1903–1907, planted rice in this area until the fatal storm of 1911 ruined the seven-mile plantation dike. He is the author of an account of rice planting in South Carolina, *Seed from Madagascar* (Univ. of North Carolina Press, 1937).

Jacksonboro, on the west bank of the Edisto River, was the place the South Carolina assembly met in 1781, when Charleston

---

SHELDON CHURCH
(Round trip about 10 miles)

Turn left on US 17. On the north side of the road is Tomotley Plantation (not open to the public). It was originally part of a 13,000-acre tract granted to Captain Edmund Bellinger, who was made landgrave of the county in 1698. The early land allotment system of the colony favored the establishment of a landed aristocracy. In each country the resident landgrave was entitled to the largest number of acres, and the landed nobility as a whole owned two-fifths of the land; the remaining three-fifths was available for sale or distribution to the people.

Further on, take a right turn toward Yemassee to the ruins of Sheldon Church. Standing brick columns in an eighteenth-century graveyard bear silent testimony to the grandeur of the Anglican edifice which once stood there on glebe lands donated by Elizabeth, widow of Landgrave Bellinger. The first building, erected in the mid-1700s, was burned by the British; the replacement was reduced to ruins by Sherman's forces during the Civil War. Sheldon was the parish church of Prince Williams Parish, one of four parishes in the Beaufort District.

Return to US 17 and turn left toward Charleston.

was occupied by the British. Just west of town on SC 64 is the *Isaac Hayne Monument.* After General Benjamin Lincoln's surrender to British General Clinton on May 12, 1780, the Continentals were kept by the British as prisoners of war, but South Carolina militiamen were freed on parole. Col. Isaac Hayne was one of those paroled. Subsequently, he was called upon as a British subject to serve in the British forces, but refused, and joined American forces. Recaptured by the British, Hayne was hanged, and became one of the martyrs of the Revolutionary War.

Just east of Jacksonboro, the road crosses the Edisto. To the east of the river, on the north side of the road, the short-lived *Stono Rebellion* was stopped on September 9, 1739. The fifty rebelling slaves were attacked by an armed contingent of whites—by tradition direct from a Presbyterian church meeting at nearby Willtown—and were soon overcome. The brief uprising had begun only that morning when slaves under the leadership of Jemmy broke into Hutcheson's store looking for small arms, killed the storekeepers, and then proceeded to kill twenty-five whites and slave informers.

The four-lane road from Jacksonboro to Charleston passes through pine country, with the remains of saw-mill piles showing the importance of lumbering in this area. At *Rantowle,* the marshes along Stono River add variety to the landscape.

Enter Charleston by the east span of the Ashley Bridge.

Charleston, South Carolina and Surrounding Areas

◬ *Day Eleven*

CHARLESTON AND VICINITY

If the traveler needs tourist information, exit left from the Ashley Bridge and go directly to the Visitors Information Center, located in the renovated train depot. Maps, brochures, a monthly calendar, hotel and restaurant information, and other tourist assistance is available.

Others may leave the bridge from the right lane and follow the Ashley River—past the municipal marina, an old rice mill, the entrance to the coast guard base, and White Point Gardens. Ahead lies East Battery, with its elevated sea walk and row of mansions on the left. On the eastern horizon are the low silhouettes of Fort Sumter, to the right, and Fort Moultrie, to the left, marking the harbor entrance. Dead ahead, in the Cooper River, lies Shute's Folly, a small island where Castle Pinckney once stood guard. Two blocks west on South Battery is Meeting Street, the heart of old Charleston.

History

Charleston began as a settlement at Albermarle Point on the Ashley River in 1670; it moved to its present site, then called Oyster Point, in 1680. Charles Towne, later Charlestown, had no municipal government, as separate from colony and state, until 1783, when it was incorporated as Charleston.

By the antebellum period, Charleston was recognized as the capital of the planter aristocracy—in manners, morals, and in politics. To abolitionists, it was the capital of the hated slaveocracy; Unionists regarded it as the hot bed of secessionism.

Almost forgotten was Charleston's hundred-year history before the Revolutionary War, and the division of Whig and Tory loyalties that made it one of the bloodiest battlegrounds in the struggle for independence. Over 140 major and minor skirmishes were fought on the soil of Revolutionary South Carolina.

In the aftermath of the Revolution, and the adoption of the Constitution in 1787, South Carolinians were nationalists and advocates of a strong federal government. John C. Calhoun, that humorless advocate of nullification and the doctrine of the concurrent majority, began his career as a War Hawk in the War of 1812. Calhoun's tomb is in the cemetery on Church Street, across from St. Phillip's.

The history of Charleston presents many contrasting images. The fun-loving Restoration monarch Charles II, on regaining the throne, gave it to the proprietors to pay off political obligations. At the outset, Charlestonians were driven to make money—by Indian trade and Indian slaving, by forest and cattle exports, and by the use of slaves to produce rice, indigo, and cotton. Not until the 1830s did trade and commerce became disreputable to the planter class of Charlestonians. This came at a time when their own consumption of luxuries and lack of production of necessities, in a country beginning to industrialize, compounded the growing economic deficit of the Charleston model despite rhetoric on economic independence.

The physical Charleston that the tourist comes to see today, the largest historic area in the United States, does not owe its existence to foresight as much as to poverty. Charleston recovered very slowly from the ravages of the Civil War, and never again had the capital to push the economic development that characterized its earlier days. It did not rebuild so much as renovate; the grand Victorian *Calhoun Mansion,* built about 1876, is atypical. The preferred Charleston style was the single-loaded house, with a porch entrance on the side facing the street.

By the 1920s Charlestonians had become self-conscious of their heritage, and in 1923 The Society for the Preservation of Spirituals, one of the first of the preservation groups, was organized. In 1931 it published *The Carolina Low Country,* an evocative series of essays and collection of spirituals. The Society for the Preservation of Old Dwellings was organized in 1920, and one of its lasting triumphs was saving the Joseph Manigault House. In 1931, Charleston adopted the first historic-zoning ordinance in the United States; and in 1944, it conducted the first citywide architectural survey: *This Is Charleston* (still in print).

In the 1940s the Historic Charleston Foundation was organized to give more financial clout to the purchase and restoration of old buildings; and much of the gentrification of Ansonborough, Charleston's first suburb, has been due to its efforts.

Perhaps nothing has done more to preserve Charleston's future by looking backward than the novel *Porgy* by Dubose Heyward, published in 1925. It was adapted for the stage; and as an opera produced in 1935, *Porgy and Bess* with music by George Gershwin. Since no black cast would play to a segregated audience in Charleston, the opera was not presented in Charleston until 1970.

For the generalist seeking an overview of history from Albermarle Point to the Spoleto Festival U.S.A., *A Short History of Charleston* (1982) by native son Robert Rosen is highly recommended. Written with wit and charm, it gives pleasure while instructing. There are two informative films about Charleston: *Dear Charleston,* shown at the Preservation Society and *Charleston Adventure,* at the Visitors Information Center.

Charleston houses reflect not only the wealth of the colonial and antebellum eras but the climate and health of this area. The immigration of Europeans and Africans to Carolina not only brought their skills but also a range of diseases unknown to the American Indian, including malaria and yellow fever. At a time when the relationship of mosquitoes and malaria was still unknown, and the sickle-cell trait of African workers not yet discovered, it was observed that in the summer months slaves sickened but died less frequently than whites. Planters withdrew in the summer to the sandhills or other places free of mosquitoes. Charleston, open to the

Charleston, South Carolina. Horse-drawn carriage. Photograph by Lynda Sterling.

Charleston, South Carolina. Women weaving sweetgrass baskets on the street. Photograph by Lynda Sterling.

Charleston, South Carolina. House across from the Battery. Photograph by Lynda Sterling.

Charleston, South Carolina. House on East Battery Street. Photograph by Lynda Sterling.

Charleston, South Carolina

1. Aiken-Rhett Mansion
2. Charleston Museum
3. New Welcome Center
4. Manigault Mansion
5. Heyward-Washington House

sea breezes, was healthier than the rice plantations, so the practice began of moving into Charleston in the summer, where there was the added advantage of sociability. Soon Charleston became the residency choice for many low-country planters, who built imposing and elaborately furnished town houses in the city and spent the winter in much simpler houses on their plantations. The Charleston "season" was April through September.

Historic Charleston, below the neck of the Ashley and Cooper rivers, is a showcase of the city landscape supported by a plantation economy. It has survived war, fire, hurricane, and earthquake to become the city's most important economic asset—history as embodied in artifact. House tours have become almost a ritual in Georgia and South Carolina as a way to raise money for historic and preservation activities. Specialized Charleston tours permit visits to many private homes otherwise seen only from the street. In March and April, the Charleston Historic Foundation sponsors an annual Festival of Houses; and in September and October, the Preservation Society sponsors House and Garden Candlelight Tours.

Charleston House Museums

Three historical homes are open to the public under the management of the Charleston Museum; another two under Historic Charleston Foundation.

The *Heyward-Washington House* was built in 1772 by Daniel Heyward, a prominent rice planter, whose son Thomas was a signer of the Declaration of Independence. The simplicity of the front-facing double house belies the richness of the interior and collection of eighteenth-century Charleston-made furniture. The "Washington" in the house name derives from the city's having rented the house in 1791 for the president's use when he made his famous southern tour.

The *Joseph Manigault House,* 1803, was designed by planter-architect Gabriel Manigault for his brother Joseph. Founder of the Manigault line in America was Pierre, who arrived in Charleston about 1695. The Manigault house is a beautiful example of Adam-style, named after architect Robert Adam who was active in London in the latter part of the eighteenth century.

The *Aiken-Rhett Mansion,* when it was built by merchant John Robinson in 1817, was a central hall, single-loaded Charleston house with south-facing piazzas and entrance on Judith Street. Interior design was the still-fashionable Adam-style neoclassic. William Aiken, Jr., who inherited the house from his father, was governor from 1844 to 1846, and made major changes in the house in two different styles: Greek Revival and Early Victorian. As a bit of contradictory fancy, at the time the interior was undergoing the Greek Revival modernization, the service courtyard was treated with Gothic Revival buidlings, including two necessaries, or privies, which are still intact. The major Victorian addition was an art gallery. Such galleries had become popular in the more ostentatious homes by the mid-nineteenth century.

Henrietta Aiken, the only child of Governor Aiken, married firebrand secessionist Robert Barnwell Rhett. After 1940, their son lived in the house with his wife Frances Dill Rhett, who gave it to the Charleston Museum in 1976.

The two house museums of the Historic Charleston Foundation are the Adam-style *Nathaniel Russell House* (1808) and the *Edmunston–Alston House* (1828), with a redecorated Greek Revival interior, original furnishings, and a superb view over the harbor.

Fort Sumter

On an artificial island in Charleston harbor, Fort Sumter deserves a visit because of its intimate association with the beginning of Civil War hostilities when Confederate fire opened on April 12, 1861, at 4:30 a.m. This brick fortification, already obsolete when the war began, had been authorized by the Congress as part of a series of coastal fortifications following the War of 1812. Construction began in 1828, but the fort was not complete or garrisoned at the outbreak of Civil War hostilities. There were only two skeleton companies and some band members, a total force of seventy-five.

The South Carolina Ordinance of Secession had been passed on December 20, 1860. On December 26, Major Robert Anderson, of Kentucky, transferred his men from Fort Moultrie to ungarrisoned Fort Sumter. He refused to yield to South Carolina demands to give up the fort. Since most U.S. forts and arsenals in the South had been taken over by state militia, the status of Fort Sumter was a question of delicate negotiation. In the meantime, Jefferson Davis had been inaugurated as president of the Confederacy on February 18. He passed on to General P. T. Beauregard, in command of Confederate forces in Charleston, the authority to set an ultimatum for the surrender of Fort Sumter. Somehow the authority to determine whether the reply of Major Anderson was satisfactory or not was delegated or assumed by two aides to Beauregard, Stephen D. Lee and Col. James Chestnut, Jr., who had resigned as senator from South Carolina when Lincoln was elected.

Fort Sumter surrendered to the Confederates on April 15, 1861, the very day Anderson had indicated that he would evacuate because of lack of provisions. Thus almost casually began a war which gave President Lincoln the excuse to call for volunteers to "save the Union." The war lasted four bloody years, the costliest in terms of combatants of any nineteenth- or twentieth-century conflict. Despite a Union siege that lasted almost that long, Sumter was never captured by Union forces. It was finally evacuated on February 17, 1865, the day that the capital, Columbia, was burned by Sherman. There was heavy destruction in the Charleston area by Southern and Union looters before order was restored.

The Ashley River Road

The Ashley River Road, SC 61, goes northwest from Charleston on the south side of the river. It is a good example of road development paralleling low-country rivers, which for many years were the principal means of transport of passengers, crops, and goods. Some plantation homes were designed symmetrically to have both a river and road front.

Three particularly interesting sites on this road are open to the public: Charles Towne Landing at Albermarle Point; the gardens at Middleton Place; and the Georgian-Palladian house Drayton Hall. *Magnolia Gardens,* which was also a Drayton property, was opened to the public in 1873; adjacent to it is the recently opened *Audubon Swamp Garden,* under Magnolia Gardens management. Old Dorchester State Park, a little farther out, is the site of a once-prosperous settlement of Massachusetts Congregationalists who chose this bluff overlooking the Ashley River in 1696. The town, which once had over 1,000 inhabitants, was abandoned in 1788. All that remains are tabby foundations of fort and houses.

Charles Towne Landing is on SC 171 about a mile north of the intersection with SC 61. At this point on the Kiawah (Ashley) River, Albermarle Point, 150 colonists landed in April 1670 to begin the first permanent English settlement in South Carolina. The Spanish regarded this English settlement as an intrusion into their territory and a violation of the treaty of 1670 in which Spain confirmed the existing possessions of Great Britain in America. The Spanish, in August 1670, sent a force to expel the intruders, but a storm dragged their ships' anchors. The Spaniards withdrew without an attempt on Charles Towne, whose palisaded earthwork was still incomplete. In 1686 and 1727, after the city had removed to the point between the Ashley and Cooper rivers, the Spanish made other attempts on Charleston, but none was successful in removing the colony. Although "setld in the very chaps of the Spaniard," the South Carolina colony flourished.

The property on which old Charles Towne had been located was never commercially developed. It was acquired by South Carolina and made into a State Park in 1970 in celebration of the state's tricentenary. The site included the eroded earthworks of the original fort and the thirty acres cleared for the growing of food crops. As a state park, Charles Towne Landing has functioned as an interpretive history center. The park was closed temporarily in 1989 after heavy damage from Hurricane Hugo, but it is gradually being restored.

Drayton Hall, approximately nine miles northwest of downtown Charleston, is a magnificent Georgian Palladian house, the only Ashley River plantation to escape the vengeance of Sherman's torchers, due to the ruse of converting it into a smallpox hospital. Built around 1740 by John Drayton, a third-generation Barbadian family who came to Carolina in 1679, the house remained in the Drayton family until the 1970s, when it was given to the state of South Carolina and the National Trust, which operates it as a museum. In architecture, Drayton Hall is unique: except for the replacement about 1800 of two of the heavy Georgian mantels with those of the lighter Adam style, the main house survived without fashionable conversions or modern conveniences, such as electricity or plumbing. Although it has lost its flankers and a garden reputed to have surpassed that of Middleton Place, the robustness of detail and symmetry of design, including the use of false doors for balance, makes this house appealing to architecture buffs. A decision was made by the National Trust to maintain the house, and not to restore or to furnish it; so no furniture or drapery competes with the elegance of its plastered ceilings or heavy cornices.

Drayton Hall was never a working plantation; it was the country estate of a royal judge (his son was to be the Revolutionary chief justice) who found it a useful site from which to manage other plantations, enjoy a circle of wealthy friends noted for their extravagance, and enjoy easy access to Charleston by the Ashley River. In addition to the Palladian porches on the road front, there is also a symmetrical front facing the river, a preferred route when transportation by river boatman in livery was much less tiring than by carriage over a muddy and uncertain road.

Middleton Place, fourteen miles northwest of Charleston, is noted for its terraced gardens overlooking the Ashley River. It was built about 1755 by Henry Middleton, president of the First Continental Congress. The founder of the South Carolina Middleton line, Edward, came to Carolina from Barbados and quickly rose to prominence in the colony. In his time Henry Middleton was one of the wealthiest men in the province; he owned 50,000 acres of land, twenty plantations, and 800 slaves. The Civil War left this family, as other prominent South Carolinians, in greatly reduced straits. The main house and flankers were burned by Sherman's troops. The gable-ended south flanker, now a house museum, is an early twentieth-century reconstruction. The property, still in the Middleton family, is vigorously managed today as one of the major "plantation" attractions of the low country.

The real glory of Middleton Place is the gardens. They were laid out so that the central axis was an extension of the hallway of the former main house, and it continued down the terraced bluff through two lakes to the river's edge. Formerly, there were extensive rice fields on the flood plain below the bluff; and the rice-mill pond not only supplemented the water vista of the garden but had a utilitarian function.

The main body of the gardens is on the north side of the entrance road. The spoke rose garden, tomb of Declaration of Independence–signer Arthur Middleton, is located here. On the river's edge is a great live oak, thirty-seven feet in circumference, estimated to be over 900 years old. The dignity of the landscaping gives an atmosphere of serenity to Middleton Place; but there is no formal record of their creation. The *South Carolina Guide* of the Works Progress Administration (1941) repeated the myth that "For 10 years 100 slaves labored to complete the 45-acre garden and 16-acre lawn." Samuel Gaillard Stoney was probably closer to fact when he wrote "With a proper plan and supervision, the work was less difficult than would seem. Many of Middleton's plantations made rice, so that his Negroes were trained in banking and ditching. Between crops there was always an idle time. Then gangs could have been brought here to shape terraces, and dig canals, without any interference with plantation work." In 1964, the Carolina Art Association issued the revised edition of the superb *Plantations of the Carolina Low Country,* with text by Stoney. This volume is notable for its superb architectural drawings as well as photographs of many houses before they were restored.

The Negro Presence in Greater Carolina

Unlike other colonies into which slaves were only gradually introduced, black slaves accompanied the first Barbadian settlers to Charles Towne in 1670. They were imported in ever-increasing numbers by planters who were already experienced with using black labor, and by 1710 had already become a majority. Not until after the Stono uprising was an effort made to curb the importation of slaves. Their numbers steadily grew by natural increase and direct importation so much that Swiss newcomer Samuel Dyssli commented in 1737 that "Carolina looks more like a negro country than like a country settled by white people."

The labor force for the raw-material–producing economy was primarily supplied by blacks held in bondage, whether the product was cattle herded in the piney woods, naval stores from the rich supply of forests, or the successfully introduced cultigens of rice, indigo, and Sea Island cotton. So important was the Carolina model of a slave-based plantation economy that the state's offshoot Georgia became a hotbed of "malcontents" who constantly clamored for the trustees to repeal the prohibition against slavery. When the trustees finally permitted slavery in 1750, there was an immediate and great increase in slavery in Georgia and the introduction of rice culture on a large scale in the coastal area. South Carolinians had been active in coastal-area land speculation, and the extension of slavery to Georgia permitted a rapid expansion of acreage devoted to rice, aided by the greater tolerance of Africans to malaria.

As a capitalized labor force, slaves received only a subsistence remuneration in food, shelter, and clothing—enough maintenance to keep them healthy and to permit them to reproduce. The planters also had their keep as children and in old age, which was a charge to production costs. But the fact that slavery was part of a profit-making capitalism also meant that the difference received from plantation commodities over labor and other costs was appropriated by the owner and not shared with the slave worker. While much of this return was reinvested in a slave economy, that is, profits were used to buy more slaves in the hope of making more profit, a substantial amount of this profit was consumed in extravagant living and the construction of great houses which were a testament to the wealth of the builder—a wealth created by slave labor.

Yet the tourist and promotional literature of Charleston is silent about the Negro contribution to this heritage, perhaps more so in the 1990s than in the 1920s when Carolinians and Georgians had become self-conscious about a genuine African culture and its survival across the years. Each year during Black History Month a plethora of articles about African Americans give vignettes about the achievements of individuals; yet there is no genuine recognition, by either white or black, of their economic importance in the creation of the South, new as well as old. As Peter Wood has commented, the failure to treat Africans as immigrants in American history, part of one of the greatest mass movements of any racial group, has helped Americans of European origin to regard black Americans as some peculiar slave misfit, a "tertium quid" in the words of W.E.B. DuBois.

Thus while the typical history of Georgia and South Carolina is full of such identifiable groups as Scots, Huguenots, Jews, Irish, Welsh, and English from specified regions of England, local history is practically devoid of blacks, as is American history except for the politics of slavery, Civil War, and Reconstruction. There is no indication that perhaps as many as 50 percent of Africans to enter what came to be the United States came via the quarantine stations on Sullivan's Island, or that moving west with the expansion of cotton into Texas over a fifty-year period went the concomitant expansion of black and white settlement, giving to the South a demographic profile as well as an economic base that would separate it from northern states.

Thus after a geographic foray across Georgia and South Carolina it is well to remember, in Charleston, that unnamed and forgotten blacks help to created a legacy which whites extol but belongs to African Americans as well. That is a chasm which neither group has yet managed to cross.

April 13, 1861

Nobody hurt, after all. How gay we were last night.

Reaction after the dread of all the slaughter we thought those dreadful cannons were making such a noise in doing.

Not even a battery worse for wear.

But the sounds of those guns makes regular meals impossible.

. . .

Not by one word or look can we detect any change in the demeanor of these negro servants. Laurence sits at our door, as sleepy and as respectful and as profoundly indifferent. So are they all. They carry it too far. You could not tell that they hear even the awful row that is going on in the bay, though it is dinning in their ears night and day. And people talk before them as if they were chairs and tables. And they make no sign. Are they stolidly stupid or wiser than we are, silent and strong, biding their time?

So tea and toast come. . . .

(Mary Chestnut's Civil War, C. Vann Woodward, ed., New Haven, 1982, p. 49)

PART THREE

Resources

⚠ Hints to the Traveler

GETTING INFORMATION

Advance preparation helps the traveler enjoy a trip. States, cities, and many small towns now have divisions of tourism and visitor services which publish many useful aids, such as maps, to the traveler. State guides tend to be more general than those published for specific cities and areas. Here are some addresses from which information may be requested.

Georgia

Georgia Department of Industry
 and Trade
Tourism Division
P.O. Box 1776
Atlanta, GA 30301

Atlanta Convention and Visitors
 Bureau
233 Peachtree Street, N.E., Suite
 2000
Atlanta, GA 30303

Brunswick and Golden Isles
 Tourist and Convention Bureau
P.O. Box 250
Brunswick, GA 31521

Savannah Coastal Refuges
U.S. Fish and Wildlife Service
P.O. Box 8487
Savannah, GA 31412-8487

Savannah Visitors Center
301 Martin Luther King
 Boulevard
Savannah, GA 31499

South Carolina

South Carolina Division of Tourism
Box 71
Columbia, SC 29202

Beaufort Chamber of Commerce
P.O. Box 910
Beaufort, SC 29901

Charleston Convention and
 Visitors Bureau
P.O. Box 975
Charleston, SC 29402

Georgetown Chamber of
 Commerce
P.O. Box 1776
Georgetown, SC 29442

Parris Island Driving Tour
Marine Corps Depot
Parris Island, SC 29905-5001

SLEEPING AND EATING

A concern of all travelers is reflected by the questions Where to sleep? Where to eat? In most places, there is a range in accommodation and food to meet most budgets—from luxurious to the simplest. The places that advertise in tourist handouts are generally in the middle and upper ranges. Cheaper accommodation is often provided by older motels, although national chains generally offer good value for the money. Increasingly popular are bed-and-breakfast inns, which often provide a regional touch in housing and furnishings, and offer a breakfast reflecting local cooking tastes. These inns tend to be somewhat higher-priced than standard accommodations.

The organizations in the previous section (except for Parris Island) provide information on accommodations and food. For a small fee, they will often make a reservation.

WHAT TO SEE AND DO

The Plantation South was written for a traveler with only a limited amount of time. The suggested eleven-day itinerary was designed to give understanding to the historical forces—economic, political, and social—which helped shape a region. Some sites, such as Hofwyl-Broadfield near Darien or Bostwick south of Athens, will not be of interest to every traveler. Additional time can easily be spent in all of the cities and many of the other stops. In Georgia, the barrier islands and the Okefenokee are barely touched. In South Carolina, the itinerary does not include historic Georgetown just north of Charleston.

WHAT TO WEAR

In general, hot, humid summers are typical of the entire region. The winter months are particularly pleasant for traveling, daytime highs are often over 50 degrees Fahrenheit in the Piedmont and 60 degrees near the coast. A layered approach is essential in the winter. A jacket and sweater combination is preferable to a heavy coat, since the mornings are chilly but the afternoons are often quite warm.

In the summer, it is common now to see entire families in shorts. While this attire is appropriate for the poolside and beach, it is not always comfortable elsewhere. Insects abound and high grass and weeds can scratch exposed, tender flesh. For any kind of trail or field walking, wear a sun hat and long-sleeved cotton shirts and pants. In shoes, fashion should give way to comfort. Both sexes should wear walking shoes and keep sandals for the beach.

TOURIST HAZARDS

Too much sun can make summer traveling unpleasant. The high sun can be extremely hot. Don't spend much time in the sun without a hat and sunscreen.

In the spring and fall, when the weather is often most pleasant, sand gnats, called "no see 'ums," abound when there is no breeze. Avon's Skin So Soft is a good preventive against gnats. For mosquitoes, red bugs, and ticks, an insect repellent with a high percentage of DEET is necessary. Spray around trouser bottoms and shoe tops, wrists, and neck.

Poison ivy, a woody vine or shrub, is found throughout eastern North America, but is especially prevalent in southeastern woods. A toxic in the plant can produce a severe skin inflammation which requires treatment by a doctor; but mild inflammation can be treated with over-the-counter medicines. Keep to marked trails and avoid plants with leaves with three leaflets. The plant's irritant can be transmitted from clothing; so after exposure, the skin and clothing should be washed.

RECREATION

In the late nineteenth century, winter hunting was one of the favorite sports in the Coastal Plain. Deer abounded then, as now, and quail and partridge shooting attracted hunters. Today the favorite recreation areas are on the barrier islands along the coast. The ocean-side pavilion, with shade and rocking chairs, has long since disappeared, but the night is now alive with bars and discos. Eating the excellent fresh seafood, in low-country style, is by some considered a part of recreation. Golf and tennis are pursued the year round. Estuarine fishing, summer and winter, has always been popular, and most resort areas also have deep-sea charter boats.

The area has many state as well as national parks. Cumberland Island, Georgia, and Hunting Island, South Carolina, are maintained in a natural state. Some parks have no overnight tourist facilities and plans must be made ahead for a day trip—some places by water.

Efforts are being made to preserve something of the coastal African-American culture. Good places to pursue this interest are in McIntosh County, Georgia, and at the Penn Center on St. Helena Island, South Carolina. Good examples of old-style basket weaving, adapted to modern usage, are sold by the weavers at the outdoor market in Charleston.

For those with a military mind, fortifications range from the simple earth breastworks to log and timber blockhouses can be visited. Fort Pulaski (Savannah) and Fort Sumter (Charleston) are fine examples of the brick coastal defense built after the War of 1812. Early twentieth-century coastal gun emplacements can be seen on Tybee Island near Savannah and on Sullivan's Island, near Charleston. A naval museum, popular with young and old, is at Heritage Point on the Cooper River opposite Charleston.

There are special festivals, museum exhibitions, and house tours throughout the year in many places. One traveler might want to plan a trip to take in one of these, such as the Spoleto Musical Festival in Charleston; another might want to avoid the crowding. When writing for information, also ask for a calendar of events.

Hunting and Fishing

Both Georgia and South Carolina require licenses to hunt or fish. The fee varies according to type of game and the duration of the license. For visitors, short-term—in contrast to seasonal—licenses are available. The licenses are usually on sale at outlets which cater to hunters and fishermen.

△ Suggested Readings

The most provocative reading to help understand the Plantation South are old travel accounts, nineteenth-century journals, and recent scholarly monographs on topics restricted in scope and time. A number of such sources are mentioned in the text. Unfortunately, many are found only in the archives of historical societies or research libraries.

If interested in reading more about the areas visited, check local bookstores and shops at various tourist attractions. Often these are the best places to find such books as the Georgia Conservancy's revised *Guide to the Georgia Coast,* H. E. Taylor Schoettle's *Field Guide to Jekyll Island,* and Robert Rosen's *A Short History of Charleston.* Margaret Mitchell's *Gone With the Wind* can be reread with an eye to social history and human geography. The author was very careful with facts and details. Also readable and well-researched are Eugenia Price's novels laid in Savannah and St. Simon's island.

The following short reading list may be of interest.

Georgia: The WPA Guide to Its Towns and Countryside, reprinted in 1990 by the University of South Carolina Press.
Originally published in 1940, this remains the best comprehensive guide to Georgia.

Journal of a Residence on a Georgia Plantation in 1838–1839 by Frances Anne Kemble, reissued in 1984 by the University of Georgia Press. This highly readable journal, which originally appeared in 1863, is a perceptive account of a winter spent at Butler Island.

Plantations of the Low Country, South Carolina 1697–1865 by N. Jane Iseley and William P. Baldin, revised edition issued by Legacy Publications in 1987.

With handsome interior and exterior photographs and a nontechnical text, this book covers plantation architecture and home furnishings.

The Souls of Black Folk by W.E.B. DuBois was reissued in 1990 by Bantam Books.

Relevant today, this 1903 work expounds in lyrical prose DuBois's belief in a liberal education for Negroes and the need for political equality.

△ Index

Adam, Robert, 169–170
agriculture, 3–14, 63–69, 110–120
Aiken, Henrietta, 170
Aiken, William, 170
Aiken-Rhett Mansion, Charleston, S.C., 170
Alapaha, Ga., 102
Albany, Ga., 92
Americus, Ga., 90
Anderson, Robert, 171
Andersonville National Historic Site, Ga., 87, 89
Anglican church, 125
Apalachee, Ga., 48
architecture, 53–54
Arfwedson, C. D., 144
Argyle Island, Ga., 150
Ashley River Road, S.C., 171–174
assets, 9
Athens, Ga., 42–46
Athens charter, 20
Atlanta, Ga., 25, 28–36; Battle of, 33
Atlanta History Center, 36
Audubon Swamp Garden, 172

Barber, Obediah, 103
barrier islands, 107–108
Bartram, William, 113
Beaufort, S.C., 157–159
Beauregard, P. T., 171
Bell, Matthew R., 58
Berrien, John MacPherson, 102
birds, 151
Blackbeard Island, Ga., 129

Bluffton, S.C., 151
Borglum, Gutzon, 39
Bostwick, John, 48
Bostwick, Ga., 48
boundaries, 19, 55
Bray, Thomas, 4
Brown, Joseph, 60
Brunswick, Ga., 109–110
Brunswick-Altamaha Canal, 121
Bryan, Andrew, 149
Bryan, Jonathan, 149
Bureau of Refugees, Freedmen, and Abandoned Lands, 58–59
Butler, Pierce, 122
Butler Island, Ga., 122–123
Butler Island Wildlife Management Area, 120
Byron, Ga., 85

Cabaniss, E. G., 86–87
Calhoun, John C., 164
Campbell, Archibald, 142
Capitol, Atlanta, Ga., 34–35
Carter, Jimmy, 25, 35, 90–91
Carter Center, Atlanta, Ga., 35–36
carving, 40
Catherines Island, Ga., 108
Cedar Shoals mill, 43–44
Charleston, S.C., 26, 163–170
Charles Towne Landing, S.C., 76, 172
Chehaw Park, Albany, Ga., 91–92
Cherokees, 73
Chestnut, James, 171
Chicora, Francisco de, 76

climate, 5
Clinton, Ga., 62
Cloud's Tower, Stone Mountain, Ga., 39
Cluskey, C. B., 53
colony, 3–4
Confederate Memorial, Stone Mountain, Ga., 39–40
commodity, 3
cotton, 5, 11–12, 63–69
Creeks, 55, 73, 75
Creek land cession, 55
crenelation, boundaries, 93
crops, 13
Cyclorama, Atlanta, Ga., 33–34

Darien, Ga., 127–128
DeBrahm, William Gerard, 103
diseases, *see* malaria
Drayton, John, 173
Drayton Hall, Charleston, S.C., 172–173
DuBois, W.E.B, 10, 176
DuPont, Samuel F., 153
Dyssli, Samuel, 175

Eagle Tavern Museum, Watkinsville, Ga., 47
Eaton, William, 51
Eatonton, Ga., 51
Ebenezer, Ga., 140–142
Ebenezer Baptist Church, Atlanta, Ga., 32
economics, 3
Edmunston-Alston House, Charleston, S.C., 170
Evans, Elijah, 48

farms, 11
First Bryan Baptist Church, Yamacrow, Ga., 149
Flagg, Wilkes, 60–61
Flagg's Chapel, Milledgeville, Ga., 61
Ford, Henry, 132

Fort, Tomlinson, 60
Fort Frederica National Monument, St. Simons Island, Ga., 124
Fort King George, Darien, Ga., 128
Fort Hawkins, Macon, Ga., 72
Fort San Marcos, Parris Island, S.C., 152
Fort Sumter, Charleston, S.C., 170–171
Fort Valley, Ga., 85
Frederica, Ga., 124–125
Freedman's Bureau, 58–59
Freedman's Grove, Ga., 132

Garden, Alexander, 159
Garden, Benjamin, 159
Gardens Corner, S.C., 159
gardening, 74
Geechee, 156
George II, 140
Georgia Agrirama, Tifton, Ga., 94
Georgia College, Milledgeville, Ga., 54
Georgia State Hospital, Milledgeville, Ga., 57
gin, saw, 5
gold, 35
"Golden Isles," Ga., 107
Grant Park, Atlanta, Ga., 33–34
Gray, James J., 61
Griswold, Samuel, 62
Grover Island, Ga., 129
Gullah, 156

Hall, Basil, 100, 115
Hancock, Walter, 40
Harper, Roland, 94
Harriet Tubman Historical and Cultural Museum, Macon, Ga., 79
Harris, Joel Chandler, 25, 36, 52
Hawkins, Benjamin, 72
Hay House, Macon, Ga., 78
Hayes, Rutherford B., 59
Hayne, Isaac, 161

headright, 17
Heritage Hall, Madison, Ga., 48, 51
Heyward, Daniel, 169
Heyward, Dubose, 165
Heyward-Washington House, Charleston, S.C., 169
Hill, Joshua, 50
Hilliard, Sam B., 115
Hilton, William, 151
Hilton Head, S.C., 151
Hodges, Robert James, 89
Hofwyl-Broadfield Plantation Historic Site, New Hope, Ga., 119–120

Indians, 73–78
indigo, 146–147
investment, 68
Isaac Hayne Monument, Jacksonboro, S.C., 161
Ivy, Hardy, 29

Jackson, Maynard, 28
Jacksonboro, S.C., 161
Jarrell, Willie Lee, 80
Jarrell Plantation Historic Site, Macon, Ga., 80–81
Jean Ribault Monument, Parris Island, S.C., 152
Jekyll Island, Ga., 26, 107–109
Jekyll Island Club, 108
Jekyll Island Orientation Center Museum, 108
Jenkins, Charles J., 58
Jerusalem Lutheran Church, Ebenezer, Ga., 140
Johnson, James, 58
Johnston, William Butler, 78
Jones, James, 61
Jones, Joseph, 115
Joseph Manigault House, Charleston, S.C., 169–170

kaolin, 85

Kemble, Frances, 122–123
Kiawah, 76
King, Martin Luther, Jr., 30–33
Komarek, E. V., 96

labor, 121, 175
Lake Sinclair, Ga., 52
land, 8, 16, 77
Langewiesche, Wolfgang, 16
Lanier, Sidney, 79, 109
Lanier Cottage, Macon, Ga., 79
Laurens, Henry, 118
Leconte-Woodmanston Plantation, Riceboro, Ga., 129
Lee, Stephen D., 171
Leesburg, Ga., 91
Lewis, George, 143
livestock, 13
Lomax, Michael, 28
lotteries, 18
Lucas, Eliza, 146
Lukeman, Augustus, 39–40
Lumpkin, Wilson, 60
Lyell, Charles, 121

McCrae, John, 41
Mackintosh, George, 110
Macon, Ga., 25, 71–73, 78–79
Madison, Ga., 48, 50–51
Madison-Morgan Cultural Center, 50
Magnolia Gardens, Charleston, S.C., 172
malaria, 88, 165
Manigault, Gabriel, 169
Marlor, John, 53
Marlor House, Milledgeville, Ga., 53
Marschner, F. J., 16
Marshall, John, 86
Marshallville, Ga., 86
Masonic Hall, Milledgeville, Ga., 53
Massee Lane Garden, Fort Valley, Ga., 85
Meade, Melinda S., 88
Melish, John, 50

Menendez, Pedro de, 76
Methodist church, 125
miasmatas, 137–138
Middleton, Arthur, 174
Middleton, Henry, 173
Middleton Place, Carleston, S.C., 173–174
Midway, Ga., 130–131
Milledge, John, 43
Milledgeville, Ga., 52–61
Minnie F. Corbitt Memorial Museum, Pearson, Ga., 102
Mitchell-Barron House, Clinton, Ga., 62
Monroe, Ga., 40–41
Montezuma, Ga., 86
Montgomery, Robert, 107
Mulberry Grove Plantation, Savannah, Ga., 65
Murray, Ellen, 154

Nahunta, Ga., 104
Nathaniel Russell House, Charleston, S.C., 170
navigability, 143
Navy Supply Corps School, Athens, Ga., 43
New Hope, Ga., 118–119
New Hope Plantation, 118–119
Nolan's Store, Bostwick, Ga., 48

oaks, live, 129
Obediah's Okefenok, Waycross, Ga., 103
Ocmulgee National Monument, Macon, Ga., 72–73
Ocmulgee Old Fields, Macon, Ga., 72–73
Oglethorpe, James Edward, 4, 109, 133–137
Oglethorpe, Ga., 86
Okefenokee Heritage Center, Waycross, Ga., 103

Okefenokee Swamp Park, Waycross, Ga., 103–104
Old Cannonball House, Macon, Ga., 79
Old Governor's Mansion, Milledgeville, Ga., 53–54
Onslow Island, Ga., 150

Paleo-Indian, 73
Parris Island, S.C., 152–153
peanuts, 13, 91
Pearson, Ga., 102
Penn Center, Port Royal, S.C., 26, 154, 156
pine barrens, 95–101
Plains, Ga., 90–91
plantations, 48, 65, 80–81, 89, 118–120, 129, 160
pochtecas, 74
Port Royal, S.C., 154
poultry, 13
public health, 139
Putnam, George Israel, 51

railroads, 29
rainfall, 7
Reconstruction, 59
Republicans, 59
revival, post–Civil War, 30
Revolution, 142, 164
Rhett, Robert Barnwell, 170
rice, 6, 110, 112
Riceboro, Ga., 129
rivers, 121
Rock Eagle, Ga., 51
Rock Eagle 4-H Center, 51
Roosevelt, Franklin D., 10
round towns, 18–21
Ruger, Thomas R., 57
Rumph, Samuel H., 86

St. Helena Island, S.C., 26, 155–156
St. Simons Island, Ga., 26, 124–125
Salzburg, 140

Santa Elena, Parris Island, S.C., 152–153
Savannah, Ga., 26, 133–146
Savannah History Museum, 139
Savannah National Wildlife Refuge, 150–153
Savannah River, 133, 143
Savannah Waterfront, 143–146
Schretter, Howard, 19
settlement pattern, 17–21
Sheldon Church, Yemassee, S.C., 160
Sherman, William T., 29, 50
slavery, 3–4, 9
slaves, 113, 174–177
Smith, Landgrave Thomas, 110
Smith House, Albany, Ga., 92
soils, 7
Southern Forest World, Waycross, Ga., 103
soybeans, 13
steamboats, 83
Stetson-Sanford House, Milledgeville, Ga., 53
Stoneman, George, 62
Stone Mountain, Ga., 25, 38–40
Stoney, Samuel Gaillard, 174
Stono Rebellion, 161
Sunbury, Ga., 131–132
Susie Agnes Hotel, Bostwick, Ga., 48
Sylvester, Ga., 93–94

Thomas, David Hurst, 108
Thronateeska Museum of History and Science, Albany, Ga., 92
Thrower, Norman J. G., 15
tide swamp, 119
Tift, Nelson, 94
Tifton, Ga., 25, 94
Toombs, Robert, 59
Towne, Laura M., 154
Tracy, Anne Clark, 78

Treaty of Paris, 76
Treaty of 1821, 72
Trebor Plantation, Andersonville, Ga., 89
Tubman, Harriet, 79
Turnwold Plantation, Eatonton, Ga., 51–52
Ty Ty, Ga., 94

Unger, Frank A., 93
University of Georgia, Athens, Ga., 42–43

vandalism, 57
Vanderhill, Burke C., 93
vegetation, 7, 100
Venable, Samuel Hoyt, 39
Vernon Square, Darien, Ga., 128
von Reck, Philip Georg Friedrich, 141

Walker, Alice, 25, 52
Walton, George, 40
war, 30, 142
War of Jenkins' Ear, 124
Ware, Nichols, 102
Watkinsville, Ga., 47
Waycross, Ga., 102–103
wheat, 12
White, John R., 44
White, Thomas W., 58
Whitney, Eli, 5, 65
wildlife, 120
Willacoochee, Ga., 102
Winberry, John, 147
Wirz, Henry, 89
Woodruff House, Macon, Ga., 79
Wren's Nest, Atlanta, Ga., 36

Yamacraw, Ga., 149

zippers, 70